How to Modify
NISSAN/DATSUN
OHC Engine
by Frank Honsowetz

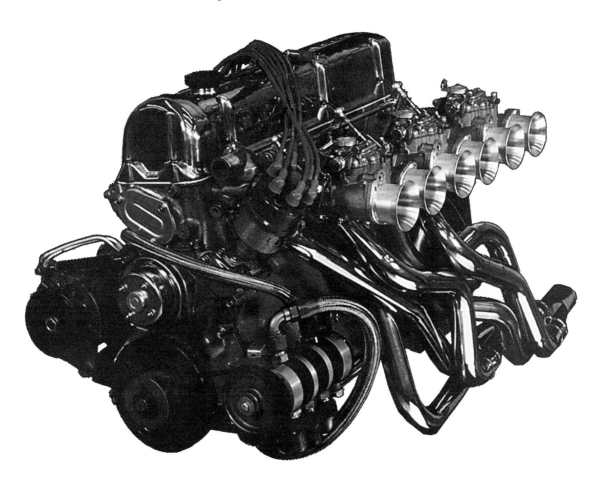

FISHER BOOKS

Contents

	Introduction	3
1	Cylinder Block	10
2	Crankshaft	19
3	Pistons & Rings	26
4	Connecting Rods	33
5	Cylinder Head	41
6	Camshaft & Valve Train	61
7	Engine Preassembly	71
8	Engine Assembly	84
9	Lubrication	96
10	Engine Electrics	101
11	Carburetion	106
12	Exhaust	126
	Suppliers Index	131
	Parts List	133
	Metric Conversions	141
	Index	143

NOTICE: The information contained in this book is true and complete to the best of our knowledge. All recommendations on parts and procedures are made without any guarantees on the part of the author or Fisher Books.

Because the quality of parts, materials and methods are beyond our control, author and publisher disclaim all liability in connection with the use of this information.

THANKS: To my wife, Toni, whose time, devotion and journalism skills made this book possible and readable. And to those of the Nissan Motorsports Department for their vast and various contributions to this book.

Publishers: Bill Fisher, Helen Fisher, Howard Fisher **Revised Edition Editor:** Bill Fisher **Original Editors:** Tom Monroe, Ron Sessions **Cover photo:** Frank Honsowetz **Photos:** Frank Honsowetz, Tom Monroe, G. Hewitt **Cover:** Gary Smith, Performance Design **Production:** Randy Schultz

© 2000 Frank Honsowetz

All rights reserved. No part of this book may be reproduced or transmitted by any means, electronic or mechanical, including photocopy, recording, or any information storage or retrieval system, without written permission from the publisher, except by a reviewer who may quote brief passages.

Printed in U.S.A.

Printing 10 9 8 7 6 5 4 3 2 1

Published by Fisher Books, LLC, 5225 W. Massingale Road, Tucson, Arizona 85743-8416
(520) 744-6110, www.fisherbooks.com

BASIC ENGINE SPECIFICATIONS

Engine	(CID/cc)	Cylinders	Bore (in./mm)	Stroke (in./mm)
L13*	79.1/1296	4	3.268/83	2.358/59.9
L14*	87.2/1428	4	3.268/83	2.598/66.0
L16	97.3/1595	4	3.268/83	2.902/73.7
L18	108.0/1770	4	3.346/85	3.071/78.0
L20A*	121.9/1998	6	3.071/78	2.744/69.7
L20B	119.1/1952	4	3.346/85	3.386/86.0
L24	145.9/2391	6	3.268/83	2.902/73.7
L26	156.3/2562	6	3.268/83	3.110/79.0
L28	168.0/2753	6	3.386/86	3.110/79.0

*Not sold in the U.S.

Library of Congress Cataloging-in-Publication Data

Honsowetz, Frank.
  How to modify your Nissan/Datsun OHC engine / by Frank Honsowetz.
    p. cm.
  Includes index.
  ISBN 1-55561-237-7
  1. Nissan automobile—Motors—Modification. 2. Datsun automobile—Motors—Modification. I. Title.

TL215.N54 H66 1999
629.25'04—dc21
99-056866

Introduction

Author in his L28-powered IMSA GTU 280Z in turn 7 at Riverside. Engine on cover powered car. Mark Yeager photo.

The 1968 510 was the first Datsun with an L-series engine imported into the United States. It had the 1595cc four-cylinder L16. A 2391cc six-cylinder version, dubbed the *L24*, was introduced in the sensational first 240Z of 1970. Later, the L-series four-cylinder engine grew to 1770cc in 1973 and finally to 1951cc in 1975.

In 1979, the original non-crossflow L-series four-cylinder head was superseded by the Naps-Z head for passenger-car use. However, 1979 and '80 truck models retained the early non-crossflow head. The Naps-Z four-cylinder engine first appeared in 1980. It had a displacement of 2187cc in 1980, growing to 2389cc in late 1983.

The six-cylinder 240Z engine was enlarged to 2562cc in 1974 for the 260Z. The engine was enlarged again to 2753cc and fuel-injected for the 280Z of 1975. From 1975 to 1984, the six-cylinder Z-car engine underwent only one significant change; an optional turbocharged version was introduced in late 1981.

1983 SCCA GT-4 National Champion, Dave Carkhuff wheels his Concord Datsun 510 at Road Atlanta.

Key features of the L-series engine are: a cast-iron, full-skirted block, non-crossflow overhead-cam (OHC) aluminum cylinder head and chain-drive camshaft. The four-cylinder block has five main bearings; the six-cylinder seven. In either case, the center main bearing incorporates the crankshaft thrust bearing.

Up top, the camshaft is supported by

---

*Naps-Z was the abbreviation for Nissan Anti-Pollution systems.

removable aluminum cam towers. The cam actuates the valves through rocker arms. The oil pump is externally mounted to the base of the cast aluminum front cover and shares a common drive spindle with the distributor assembly.

RACING HISTORY

The Datsun 510 and 240Z were immediate successes in sports-car road racing. In the United States, Datsun supported the winning efforts of Brock Racing Enterprises (BRE) and Bob Sharp Racing. In 1970, BRE won the first of 10 consecutive "Z-Car" Sports Car Club of America (SCCA) National Championships. The 510 was equally successful. Bob Sharp won National Championships in 1971 and 1972, driving a 510.

The highlight of the 510's racing career was two consecutive SCCA 2.5 Trans-Am Championships in 1971 and '72. In doing so, the BRE 510s beat European performance-image sedans such as BMW and Alfa Romeo. During the same period, Bob Sharp Racing began a winning record in L-series powered Datsun sedans and Z-cars that continued for 16 years.

Throughout the mid and late '70s, Bob Sharp continued racing and winning with

BASIC ENGINE SPECIFICATIONS

Engine	Displacement cc/inch3	Bore mm/in.	Stroke mm/in.
L16 (four-cylinder)	1595/97.3	83/3.268	73.7/2.902
L18 (four-cylinder)	1770/108.0	85/3.346	78.0/3.071
L20B/Z (four-cylinder)	1951/119.1	85/3.346	86.0/3.386
Z22 (four-cylinder)	2187/133.4	87/3.425	92.0/3.622
Z24 (four-cylinder)	2389/145.7	89/3.504	96.0/3.779
L24 (six-cylinder)	2391/145.9	83/3.268	73.7/2.902
L26 (six-cylinder)	2562/156.3	83/3.268	79.0/3.110
L28 (six-cylinder)	2753/168.0	86/3.386	79.0/3.110
L16 + 1mm/0.040-in. overbore	1636/99.8	84/3.307	73.7/2.902
L18 + 1mm/0.040-in. overbore	1816/110.8	86/3.386	78.0/3.071
L20B + 1mm/0.040-in. overbore	2005/122.3	86/3.386	86.0/3.386
Z22 + 1mm/0.040-in. overbore	2230/136.1	88/3.464	92.0/3.622
Z24 (4) + 1mm/0.040-in. overbore	2438/148.7	90/3.543	96.0/3.779
L24 (6) + 1mm/0.040-in. overbore	2454/149.8	84/3.307	73.7/2.902
L26 + 1mm/0.040-in. overbore	2630/160.5	84/3.307	79.0/3.110
L28 + 1mm/0.040-in. overbore	2824/172.3	87/3.425	79.0/3.110

BORE/STROKE COMBINATIONS

Bore/Stroke	Displacement cc/in.3	Bore mm/in.	Stroke mm/in.
L16 Bore/L18 Stroke	1686/102.9	83/3.268	78.0/3.071
+ 1 mm/0.040-in. overbore	1731/105.6	84/3.307	78.0/3.071
L18 Bore/L16 Stroke	1672/102.0	85/3.346	73.7/2.902
+ 1 mm/0.040-in. overbore	1716/104.7	86/3.386	73.7/2.902
L20B/ZBore/L22Z Stroke	2086/127.2	85/3.346	92.0/3.622
+ 1 mm/0.040-in. overbore	2141/130.6	86/3.386	92.0/3.622
Z22 Bore/L20B/Z Stroke	2052/125.2	87/3.425	86.0/3.386
+ 1 mm/0.040-in. overbore	2089/127.4	88/3.464	86.0/3.386
Z24(4) Bore/L22Z Stroke	2287/139.6	89/3.504	92.0/3.622
+ 1 mm/0.040-in. overbore	2335/142.3	90/3.543	92.0/3.622
Z24(4) Bore/L20B/Z Stroke	2142/130.7	89/3.504	86.0/3.386
+ 1 mm/0.040-in, overbore	2186/133.4	90/3.543	86.0/3.386
L28 Bore/L20A Stroke	2434/148.5	86/3.386	69.7/2.744
+ 1 mm/0.040-in. overbore	2492/152.0	87/3.425	69.7/2.744
L28 Bore/LD28 Stroke	2901/177.0	86/3.386	83.0/3.268
+ 1 mm/0.040-in. overbore	2967/181.2	87/3.425	83.0/3.268
+ 3 mm/0.118-in. overbore (Nissan Motorsports "Big Bore Kit" 99996-28BBK)	3098/189.0	89/3.504	83.0/3.267
L24/26 Bore/LD28 Stroke	2699/164.7	83/3.268	83.0/3.268
+ 1 mm/0.040-in. overbore	2766/168.7	84/3.307	83.0/3.268
L28 Bore/L24 Stroke	2575/157.1	86/3.386	73.7/2.902
+ 1 mm/0.040-in. overbore	2636/160.8	87/3.425	73.7/2.902

Sectioned L28 shows pistons, rods and cylinder-head layout of L-series Nissan/Datsun engine.

Original BRE 510 car-number 46 as it exists today: Car is used for shows and displays.

From left to right John Morton, Bob Sharp and Dan Parkinson lead start of 1971 C-Production National Championship runoff. John Morton emerged the victor.

Datsuns in both SCCA and International Motor Sports Association (IMSA) events. Several other Datsun teams were successful as well, including those of Walt Maas and Frank Leary with FAR Performance.

In 1978, Dick Roberts, Datsun Competition Department Manager—Nissan Motorsports Department up to September 1985—promoted Don Devendorf and the Electramotive team from racing small Datsun sedans to a Z-car for competition in the IMSA GTU category—Grand Touring under 2.5 liters. Devendorf and his partner, John Knepp, then accelerated the development of the L-series engine.

Bob Sharp Racing and Electramotive both campaigned GTU Z-cars in the IMSA series in 1979. Devendorf's Electramotive team won the championship that year. Sharp's team, with Paul Newman driving, also competed in SCCA events, winning the C-Production (C/P) National Championship the same year.

From 1971 to 1981, Datsun sedans powered by L-series four-cylinder engines won 10 National Championships. Dave Frellsen and Bob Sharp Racing were

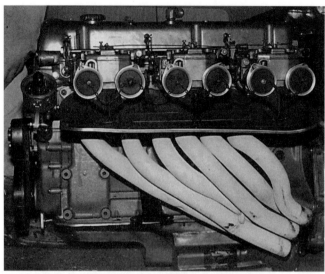

As of 1985, Gerry Mason Sr. and Jr. were competitive in SCCA GT-2 category. Their Z-car, which was built by BRE, uses many of the original engine components.

In 1983, Jim Fitzgerald competed in SCCA National Championships with a 280ZX powered by Electramotive-built L28.

L28 with Mikuni 50 PHH carburetor powered author's IMSA GTU Z-car.

Bill Long's SCCA GT-3 spec L20B. Engine uses early FIA Group-2 cylinder head and 50mm carburetors.

responsible for eight of the 10 wins.

In 1981, Datsun introduced the 280ZX Turbo and, during the 1981—83 race seasons, moved to higher racing categories. As a result, the Datsun 280ZX Turbo competed against V8-powered Detroit pony cars and exotic European sports cars.

In 1982, Devendorf won the IMSA GTO (Grand Touring over 2.5 liter) Championship in an Electramotive-prepped Turbo-ZX. During the same year, Gene Crowe of Bob Sharp Racing developed a Turbo-ZX that dominated the SCCA GT-1 category. Sharp's car, piloted by Paul Newman, was entered in SCCA Trans-Am races and in 1982 at Brainerd Raceway, he won handily. Another tough competitor, Frank Leary, also built a Turbo-ZX and entered it in selected IMSA and SCCA events.

In the early '80s, teams such as Carney/Davenport, Logan Blackburn, and Leitzinger Racing also took advantage of Electramotive's engine technology to win numerous IMSA races. In 1982, the Don Preston prepared Carney/Davenport 280ZX finished fourth overall in the 24 Hours of Daytona. In 1984, Morris Clement upheld Datsun's winning image at the SCCA National Championships in his 2.8-liter six-cylinder L-series-powered 280ZX.

With turbocharged race cars becoming more dominant, Don Preston developed a turbo 1.8-liter four-cylinder 200SX for the

Spencer Low compete with 2.4-liter-powered Class 7S off-road truck. Look carefully and you can see engine buried under all those struts and shock absorbers. Engine develops excellent power over broad rpm range.

Bob Sharp Racing built 280ZX Turbo that Paul Newman drove in 1983 Trans-Am Series.

Complex ducting is required for turbocharged Bob Sharp Racing 280ZX Trans-Am car. From left to right, ducting is for: (1) brakes, (2) radiator, (3) turbocharger intercooler and (3) induction air intake with (4) air cleaner.

Gene Crowe relocated distributor to front of rocker-arm cover, where it is driven off nose of cam. Bosch fuel-injection pump is mounted in original distributor location.

Carney/Davenport Team. George Alderman also joined the "get a turbo, go faster" group with a 2.0-liter six-cylinder (L20A) turbocharged 280ZX.

Datsun continued to win National Championships with L-series, four-cylinder engines. Mike Rickman, Rob Dyson and Bill Coykendall all won with L-series engines in 200SXs. Some of the engines for these cars were built by John Caldwell, a long-time Nissan engine builder whose initial association with Datsun engines dates back to BRE, as do other successful Nissan/Datsun engine builders Floyd Link and Electramotive's John Knepp. The longtime Nissan/Datsun race history was furthered when Dave Carkhuff won a 1983 National Championship in a Datsun 510 that was more than 10 years old. Bob Sharp's son, Scott, again demonstrated the performance of early Z's with L6 engines by winning the 1986 GT-2 National Championship with one of his father's earliest Z cars. In 1998 Jim Goughary won the GT-2 National Championship with a L6 powered 1995 model Nissan 300ZX. The current total of SCCA National Championships won by L-series engine now stands at an amazing 34.

This Book—I organized this book around each major component or system. For example, I start with the foundation of the engine, the block, then proceed to discuss all other engine components or systems and how to choose, apply and prepare them for specific high-performance applications. Once you've chosen and prepared the engine components, I show how to preassemble an engine to check critical clearances, fits and

Bob Sharp Racing built induction system for 2.5-liter 0.060-in. overbore turbocharged L24. Valve spring at front of plenum chamber is part of emergency pressure-relief valve.

relationships between certain components. I then go through final engine assembly step by step.

As a companion to this book, get Fisher Books' *How to Rebuild Your Nissan/Datsun OHC Engine*, Nissan part number 99996-M8013. Where I've assumed a high degree of prior knowledge, the companion rebuild book provides great detail on the rebuilding of the L-series Nissan/Datsun engine. Coverage includes troubleshooting, engine removal, teardown, inspection and reconditioning, assembly and installation.

Buried under all the hardware is a turbocharged L28. Engine is mounted on Electramotive's dynamometer and outfitted as it would be in Don Devendorf's IMSA GTO race car.

Introduction 9

Devendorf's own electronic fuel-injection system enhanced performance of his IMSA GTO racecar.

Turbocharged L28 was installed in Devendorf's 280ZX at angle shown. Engine plate mounted to rear face of engine block secures engine to chassis.

Morris Clement's crew attached lifting hoist to remove practice engine. "Good" engine went in its place for running 1984 SCCA National Championships, which he won.

CHAPTER ONE
Cylinder Block

Morris Clement en route to 1984 SCCA GT-2 National Championship behind wheel of his 280ZX.

Cut-away engine reveals non-Siamesed cylinder walls (arrow) of early L28 block.

Unlike most other manufacturers' engine blocks, all Nissan/Datsun blocks are suitable for use in competition. This is fortunate because Nissan does not produce heavy-duty or racing engine blocks. There's no need to.

All L-series engines share the same basic block design except for that used in the 280ZX Turbo. The first L28 Turbo blocks have the front three and rear three cylinders *Siamesed*. Webs are cast between cylinders 1 and 2, 2 and 3, 4 and 5, and 5 and 6. Although this improved block rigidity, it was thought that the webs inhibited coolant flow around the Siamesed cylinders.

To improve coolant circulation, a casting change was made. On the L28 Turbo block 11010-P9080, introduced in the 280ZX Turbo in December, 1980, and non-turbo L28s in July, 1981, slots were put in the Siamesed cores to provide coolant passages between the cylinders. Curiously enough, there were no incidents of piston-bore scuffing with the earlier L28 unslotted Siamesed blocks which would have indicated the need to use the later block for racing. However, the turbo block was used exclusively on Don Devendorf's Electramotive 600HP 280ZX Turbo L28. It was also used in the Bob Sharp Racing 280ZX Turbo.

Over the years there have been numerous changes to the size and location of

Front cover of L28 uses one 8mm securing bolt at top on each side (arrow). L16, L24 and L26 use same setup.

The 19.6mm (0.77-in.) taller L20B, L20Z and L22Z blocks use two closely spaced 8mm bolts at top of front cover on each side (arrows).

water-passage holes at the head-gasket surface. There also have been several head-gasket water-passage arrangements. None of these factors appear to be critical as to which cylinder block is best for racing.

STREET/HOT-ROD PREP

The level of preparation you need to perform on a cylinder block depends on the specific application. When building an engine for street or hot-rod use, minor work is all that's necessary. The block should be cleaned—hot-tanking is preferable—particularly if the engine was full of baked-on oil and sludge. The *core plugs*—frequently called *freeze plugs*—must be removed and replaced with new ones. Check the cleanliness and condition of the threaded holes. Even though they rarely warp, check the head-gasket surface—*deck*—for straightness. The deck should be very clean and have no signs of corrosion that may have resulted from a leaking head gasket.

CYLINDER-BLOCK SPECIFICATIONS
mm (in.)

Engine	Cylinder Bore*	Overall Height#	Main-Bearing Housing Bore Diameter
L13	83 (3.268)	265.9 (10.469)	58.66-58.67 (2.3094-2.3099)
L14	83 (3.268)	265.9 (10.469)	58.66-58.67 (2.3094-2.3099)
L16	83 (3.268)	265.9 (10.469)	58.66-58.67 (2.3094-2.3099)
L18	85 (3.346)	265.9 (10.469)	58.66-58.67 (2.3094-2.3099)
L20B	85 (3.346)	285.5 (11.240)	63.61-63.68 (2.5043-2.5073)
L20A	78 (3.071)	265.9 (10.469)	58.66-58.67 (2.3094-2.3099)
Z20	85 (3.364)	285.5 (11.24)	63.61-63.68 (2.5043-2.5073)
Z22	87 (3.425)	285.5 (11.24)	63.61-63.68 (2.5043-2.5073)
Z24	89 (3.504)	305.5 (12.03)	63.61-63.68 (2.5043-2.5073)
L24	83 (3.268)	265.9 (10.469)	58.66-58.67 (2.3094-2.3099)
L26	83 (3.268)	265.9 (10.469)	58.66-58.67 (2.3094-2.3099)
L28	86 (3.386)	265.9 (10.469)	58.66-58.67 (2-3094-2.3099)

*Typically, a 2.0mm (0.080-in.) maximum overbore will be OK without sonic testing.
#Bottom of pan rails to deck surface.

Note wide spacing of top front-cover bolts. LZ24 block is 20mm (0.788-in.) taller than L20B block. David Day's LZ24 front cover has been modified to accept early SSS race head.

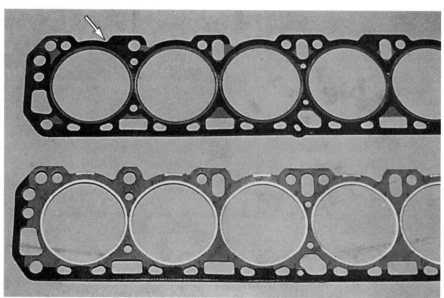

Late L28-Turbo block requires unique head gasket (top). Gasket seals additional-water passage (arrow) just rear of head-bolt hole between cylinders 5 and 6 on manifold side.

Late L22Z block has a relatively good interior finish compared to some blocks. Note how number-4 bearing-shell oil hole (arrow) doesn't quite line up with oil passage in block.

Rebore—Nissan/Datsun cylinder blocks are tough. Consequently, an overbore is rarely required. However, if an overbore is necessary for whatever reason, always take the block *with its main caps installed* and the new pistons and rings to the machinist who's going to do the rebore. He'll need the pistons to size the bores for correct piston-to-bore clearance. The piston-ring material will determine the cylinder-bore hone finish.

Get it Clean—Cylinder-block cleanliness, especially in the bores, is very important. During final cleaning, start by washing the block with clean solvent. Then scrub the bores with soap and water using a non-metallic brush.

After rinsing with clean water and air blow drying, apply a water-dispersant oil, such a WD-40® to prevent rust. While applying the oil, wipe the cylinder bores with a clean low lint white paper towel to check for *absolute* cleanliness.

If the towel comes out dirty, scrub the block again and recheck for cleanliness. Finally, paint the block exterior and store for assembly.

RACE PREP

If you're going to transform a stock block into a race block, check its general condition more closely than if you were doing a typical rebuild. Measure cylinder bores and main-bearing bores. Look for damaged surfaces, pulled threads in screw holes and any signs of cracks. Continue inspecting for any faults or potential problems during the entire cleaning and prep process. It's always easier and cheaper to correct a problem before an engine is run than it is to correct the resulting damage afterward.

Clean Block—If you haven't already done so, remove the core plugs and clean the block. You can do the initial cleaning in solvent, providing the block is relatively clean. Or, if the oil passages and water jackets are the least bit dirty, have the block hot-tanked.

After the solvent bath or hot-tanking, clean all gasket surfaces and main-bearing bores with fine sandpaper. Although it's highly unlikely that you'll find a

Cylinder Block

Tap each end of main oil gallery for 1/4- or 3/8-in. socket-head pipe plug. Afterward, flush gallery to remove all traces of metal cuttings. Tom Monroe photo.

Front of main oil gallery is sealed with pipe plug. Plug was ground to install flush with front face of block so it won't interfere with front-cover sealing.

problem, check the head-gasket surface for straightness and the main-bearing bores for roundness. You'll need a precision straightedge and feeler gages, and a dial bore gage or telescoping gage and micrometer.

Check the deck surface with feeler gages and a straightedge. Position the straightedge diagonally both ways and lengthwise in several locations from side to side on the deck surface. You shouldn't be able to fit a 0.002-in. (0.05mm) feeler gage between the straightedge and block. If you can, have the deck resurfaced.

CAUTION: Don't be tempted to use a sander, especially an "airboard" body sander, to clean the gasket surfaces of either the block or cylinder head. A sander will not stay square with the surface and may round off the edges so the head gasket can never seal correctly.

To check the main-bearing bores, install the caps and torque the bolts to 40 ft-lb. Make sure the bolt threads and those in the block are cleaned and oiled first. Now you can check bearing-bore roundness. Start with the front or rear main-bearing bore. If you're using a dial bore gage, position its plunger parallel to the vertical centerline of the block and then zero the dial. Rotate the gage 45° to each side and check each reading. Or, measure with your telescoping gage set in these three positions and compare the measurements.

In either case, measurements shouldn't vary by more than 0.001 in. (0.025mm). If any does, have the block *align-honed.*

The edges of machined surfaces should be radiused and cast surfaces of the block interior smoothed with a high-speed die grinder and rotary stones. Smoothing the block interior removes casting flash and any remaining particles from the sand-casting mold. This should be done because high-speed vibrations experienced under race conditions can loosen flash and casting sand particles. Radiusing also reduces the chances of stress cracks in the block, and helps prevent cuts when handling the block.

Oil-Gallery Plugs—Remove the press-fit main oil-gallery plugs at the front and rear of the block. Once out, drill and tap the holes for a 1/4- or 3/8-in. National Pipe Thread (NPT) plug. Installing pipe plugs accomplishes two things. It allows you to clean out the oil galleries and will ease subsequent oil-gallery cleanings.

The front plug must be shortened so it doesn't close off the number-1 main-bearing oil passage and at the same time will fit flush with the front face of the block. The plug must not interfere with the fit of the front cover. Use internal-hex Allen-type, steel NPT plugs at both ends of the gallery.

The oil-pressure-sender hole in the side of the block adjacent to the oil filter

Oil-pressure tap at right of oil-filter boss has dash-3 male fitting for line that goes to gage. Relief valve has been removed and plugged with 3/8-in. pipe plug (arrow). A 1/2-in. NPT/dash-10 male fitting is installed in place of oil-filter nipple.

can be easily modified to accept a 1/8-in. NPT fitting. Do this by chasing—running an 1/8-in. NPT tap through-the existing hole. This will allow you to install a 1/8-in. NPT dash-3 or -4 (3/16- or 1/4-in. ID) male union for the oil-pressure-gauge line.

The use of a high-pressure racing type oil filter and remote filter adapter requires plugging the oil-filter relief-valve passage.

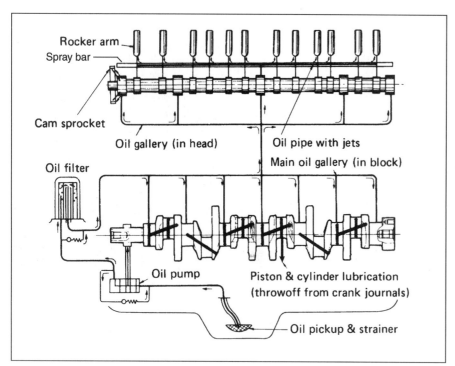

Production oiling system of six-cylinder L-series engines produced before March 1977 use a spray-bar to lubricate cam lobes and rocker arms. Six-cylinder engines built after March '77 use internal camshaft oiling for cam lobes and rocker arms. Each main bearing feeds one connecting-rod bearing. Drawing courtesy Nissan.

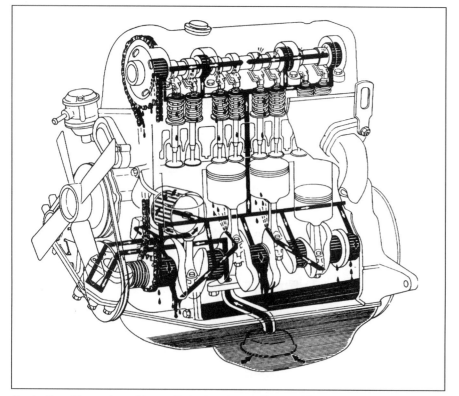

Production oiling system of four-cylinder L-series engines use direct oiling for cam lobes and rockers arms. Note that the number 2 and 4 main bearings feed two connecting-rod bearings each. Drawing courtesy Nissan.

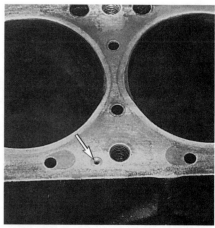

Oil jet in block deck (arrow) supplies oil to cylinder head. Relatively small hole indicates how little oil is required for valve-train lubrication.

The valve is next to the oil-filter nipple. Gain access to the passage by removing the complete relief-valve assembly, including its housing sleeve that's pressed about 1 in. deep in the passage. A soft core plug can be used, but it's better to drill and tap the hole for a 3/8-in. NPT plug.

Install all plugs with epoxy to reduce the possibility of oil leakage.

Threaded Holes—To ensure correct bolt torquing, chase all threaded holes with the correct tap. Coat the tap with light oil for lubrication. *Chamfer* all threaded holes to reduce the possibility of the top threads pulling, page 77. During the final cleaning process, use a gunbore brush to clean all threaded holes.

Chain Guide—Although uncommon, there have been incidents of the slack-side timing-chain guide top bolt being sheared from violent chain action. The *slack-side* chain guide is the curved one on the same side of the chain as the tensioner. To eliminate any possibility of the bolt shearing, increase bolt size to 8mm. or 5/16 in. Also, use a Grade-8 bolt.

To install the larger bolt, drill out the existing bolt hole and retap it to the correct thread size and pitch. Enlarge the corresponding hole in the chain guide to 8mm or 5/16 in.

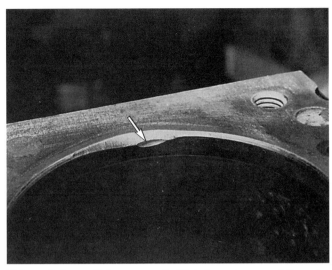
Valve relief wasn't deep enough. Nick (arrow) was put there by exhaust valve. Valve was also bent. Problem could've been found if engine was mocked up prior to final assembly, see page 71.

Besides sealing cast surface, painted block interior aids cleaning of block during freshening or rebuilding. It also helps the oil to run off of the surfaces faster.

Camshaft Oiling—Six-cylinder engines use either a spray bar or *internal*-drilled cam for camshaft-lobe/rocker-arm lubrication. For internal lubrication, the cam is drilled lengthwise. Each lobe is drilled normal (90°) to its wiping surface.

All four-cylinder engines use internal cam-and-lobe lubrication. On four-cylinder race engines, an external-cam-cover-mounted-spray bar should be installed, page 100. Oil supply for the external spray bar is from the oil-pressure sender hole on the right side of the block.

On all production L-series engines oil supply to the cylinder head is regulated by an oil jet—*restriction*. This jet is installed in the oil gallery in the block deck between the center two cylinders on the oil-filter side of the block.

The question of whether to enlarge this oil jet generates many differing opinions. However, this question becomes clearer when the situation is put into context. Engines with high valve-spring "pressure"—*load*—and high lift, radical-profile cams are harder on cam lobes and rocker-arm surfaces. In such cases, additional oil to the valve train may be desirable so long as it doesn't deprive the bottom end of lubrication.

The standard oil-jet size is 0.0787 in. (2.0mm). Enlarging the hole to 0.125 in. (3.17mm) increases jet area 2.5 times. Such an increase will not affect the bottom-end oil supply if the oil pump provides adequate volume.

Bottom-End Oiling—On six-cylinder engines, the number-1 main journal feeds oil to the number-1 connecting-rod journal, number-2 main feeds the number-2 rod journal and so on to the rear of the engine. The rear main doesn't supply oil to a rod journal. This oil flow through the block and crankshaft is adequate for racing applications.

The situation with four-cylinder engines is different. Rather than one rod journal receiving lubrication from one main journal, the number-2 and -4 main journals provide oil to all four connecting-rod journals. Therefore, oil for the four connecting-rod journals is restricted because two rod bearings must share oil from one main bearing.

To help ensure that the connecting rods in your four-cylinder engine receive adequate lubrication during high rpm operation, enlarge the number-2 and -4 main-bearing oil-supply holes. Standard size is 6mm (0.237 in.). Enlarge these oil holes to 8mm (0.313 in.). This is easily done with a standard slow-speed 1/2-in. electric drill with a 5/16-in. drill bit. Drill into the oil passage at the main-bearing bore. Stop drilling when you intersect the main front-to-rear oil gallery. After drilling both passages, deburr the edges and flush out the oil gallery to remove all metal particles.

Valve Reliefs—When 35mm exhaust valves are used on engines with 83—85mm bores and high-lift cams, the bores must be relieved, or *notched*, in the area of the exhaust valves. A relief at the top of each bore *unshrouds* the exhaust valve during high lift, improving air/fuel-mixture flow and, consequently, power.

To layout the valve reliefs, use the head gasket as a pattern to determine where the exhaust-valve reliefs go and how much relieving can be done. Coat the area of the intended valve relief on the deck with machinist's blue. With the head gasket in place over the cylinder-head dowels, scribe the shape of the reliefs. To cut the reliefs, use your die grinder fitted with a rotary stone. Do not grind the block so the underside of the gasket will be exposed to combustion heat. Only the *edge* of the head gasket should be exposed.

Also, caution must be used to ensure that the relief does not extend down the

O-ring grooves are positioned at OD of head-gasket fire rings. Except between number-3 and -4 cylinders, O-ring grooves intersect on this L28 block. If you were wondering, piston-dome damage occurred when valve head broke off at high rpm.

O-ring groove being cut with Iskenderian's Groove-O-Matic portable cutter. After tool is set up in bore, it's rotated to cut a 0.038-in.-wide X 0.030-0.034-in.-deep groove. Tool can be rented or purchased.

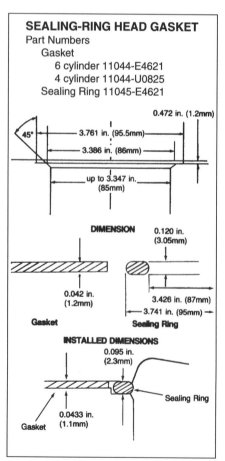

To convert from conventional-type head gasket to sealing rings, block must be machined at top of each bore as shown.

bore and infringe on the surface wiped by the top piston ring at TDC. To prevent this from happening, leave a 0.040-in. (1mm) safety margin from the top edge of top ring to the bottom of the exhaust-valve relief.

Note: Gaskets that are perfectly round at the bores can't be used with blocks that have exhaust-valve reliefs or vice versa. Also, big overbores—87mm (3.425 in.) and over—leave little room for "conventional" reliefs. However, any relieving that's possible will help.

When you finish putting valve reliefs in the tops of the bores, turn the block over and inspect the bottom end of the bores. Treatment to this end varies with engine builders. Nissan Motor Co. Ltd,'s *Competition Preparation Manual* for the L20B engine specifies a 1mm (0.040-in.) radius at the bottom edge of each bore. For your purposes, either do this or, at least, remove the sharp edge and any burrs from the bottom of each bore after boring and/or honing.

Interior Painting—Although time-consuming, painting the interior of the cylinder block is beneficial. As with any cast-iron block, Nissan/Datsun blocks have their fair share of casting-sand residue.

To prevent this sand from breaking loose and getting into your engine's lubrication system, start by removing it. If you haven't already done so, polishing the block interior, page 13, will break loose most residual casting sand. Next, regardless of whether or not you've polished the interior, clean the block thoroughly to remove sand, abrasive grit, dirt and metal particles. Once clean and dry, paint the block interior to seal the cast surfaces. Use a small brush and lots of care to apply Rustoleum® Oxide Red Primer or General Electric Glyptol®.

Head-Gasket O-Rings & Sealing Rings—Head-gasket sealing can be a problem with high-compression racing engines. To reduce or eliminate such a problem, *O-ring* the block deck surface with four or six 0.040-in. wire rings. These O-rings are installed in 0.030—0.032-in.-deep grooves in the deck surface around each bore. Each O-ring groove must be cut so it is concentric with the bore and matches the head-gasket fire-ring OD. O-ring grooves must not intersect any water passages or bolt holes.

Circular O-ring grooves can be cut

Cylinder Block

> ## SEALING RINGS
>
> Nissan Motorsports' head-gasket/sealing-ring setup ensures maximum combustion-chamber sealing. Such a setup is mandatory for all turbocharged racing engines. However, sealing rings have two disadvantages: high cost and special installation. But, if the application requires it, these disadvantages are outweighed by the consequences of not using sealing rings.
>
> Sealing rings and special head gaskets are available for four- and six-cylinder L-series engines. Rings are also available for two bore-size ranges:
>
> | Application | Nissan Part Number |
> |---|---|
> | Six-cylinder head gasket | 11044-E4621 |
> | Four-cylinder head gasket | 11044-U0825 |
> | Sealing ring for bore sizes up to 3.425 in. (87mm) | 11045-E4621 |
> | Sealing ring for 3.425—3.496-in.. (87—88.8mm)) bore sizes. | 11045-N3120 |
>
> The inside and outside diameters are the only differences between the two available sealing rings.
>
> The sealing rings are approximately twice as thick as the accompanying gasket. Consequently, the top of the cylinder block must be counterbored to accept the sealing rings. The installed or compressed thickness of the special head gasket is 0.0433 in. (1.1mm). This results in an installed or compressed sealing-ring thickness of 0.0905 in. (2.3mm).
>
> The counterbore must be machined to an exact depth of 0.04725 in. (1.2mm) and centered exactly on the cylinder bore. The counterbore OD must be 0.020-in. (0.5mm) larger than the sealing-ring OD. Sealing-ring ODs are 3.76 in. (95.5mm) and 3.78 in. (96mm), respectively. Therefore, the 3.425-in. (87mm) bore-size ring requires a 3.76-in. (95.5mm) OD counterbore; the 3.49-in. (88.8mm) bore-size ring requires a 3.78-in. (96mm) OD counterbore.
>
> Sealing-ring and gasket installation is relatively straightforward after certain critical dimensions have been confirmed. For example, sealing-ring ID must be 0.020-in. (0.51mm) larger than piston OD. If there's not adequate sealing ring-to-piston clearance, the top edge of the piston must be machined to clear the ring. For this reason, Nissan Motorsports has a step machined on some of its forged pistons to give adequate sealing-ring clearance.
>
> The standard thickness of a new sealing ring is 0.120 in. (3.05mm). Sealing rings are reusable, but minimum thickness is 0.102 in. (2.6mm). The maximum thickness variation in a set of sealing rings is 0.004 in. (0.1mm). If a ring is thinner than the minimum thickness, or one or two rings are a different thickness than the rest, they must be replaced with new pieces. Generally, if one ring is not to specification, the entire sealing-ring set must be replaced to ensure that all the rings match so maximum sealing is obtained.
>
> To install a sealing-ring head gasket, position each sealing ring seam-down in its counterbore. Apply a sealer such as High-Tack to the head gasket. Install the two gasket/head alignment dowels, one at the front and one near the rear. Position the gasket to the dowels and around the rings. Be sure the rings stay in position. Before installing the head, make certain the gasket surface on the head is clean. Install the head on the block, making sure the sealing rings don't shift out of position.
>
> Tighten the head bolts in the normal manner, from the center out, in 10 ft-lb increments to a final torque of 65—70 ft-lb. After initial warmup, let the engine cool, then retorque the head bolts 65—70 ft-lb. If there are any traces of water seepage around the head gasket, add a stop-leak additive to the cooling system.

with a tool such as that offered by Iskenderian Racing Cams. This tool, which registers in the cylinder bore, is portable, relatively inexpensive and reasonably easy to use. Although considerably more expensive, circular grooves can also be cut in a mill. A mill can also cut non-circular or asymmetrical grooves that follow a tracer pattern. Some engine builders prefer grooves that follow the pattern of the head-gasket fire ring.

An O-ring groove for 0.040-in. wire should be 0.038 in. wide and 0.030—0.034 in. deep. I prefer copper wire, although stainless steel is used by some engine builders.

Use care when cutting the wire to length to ensure proper *butting* of the O-ring in its groove. It should not gap or overlap at the ends. Where the ends are placed is not critical.

When installing an O-ring, use a plastic hammer—not steel or even lead. This ensures that the O-ring wire will not be damaged during installation. O-ring wire can be reused once or twice so long as it's not deformed.

Sealing rings and cylinder-head gaskets designed for use with sealing rings are available from Nissan Motorsports for L-series four- and six-cylinder engines. The sealing rings offered by Nissan Motorsports for L-series engines are the Cooper type, which look like gas-filled rings but are not. Cooper rings feature a Z-shape cross-section beam structure that provides exceptional resilience. They are available in two bore-ID-sizes and will fit all L-series engines.

Sealing rings are used successfully on all turbocharged four- and six-cylinder racing engines. They provide the ultimate compression seal, but do have some disadvantages. In some cases, water seepage can occur. And sealing rings and special gaskets are expensive. Also, the top of each bore must be counterbored for registering the sealing rings.

The sealing rings must not overhang into the bores and interfere with the pistons. Otherwise, the pistons must be machined with a step around their top edges to clear the sealing rings.

Refer to the drawing and installation instructions when preparing your engine for sealing rings.

Dry-Sump Applications—On dry-sump-lubrication systems, covered in Chapter 9, you can eliminate the remote oil-system adapter that installs in place of the oil filter. Do this by drilling and threading the filter-nipple hole to 1/2-in. NPT. Install a 1/2-in. NPT-to-dash-10

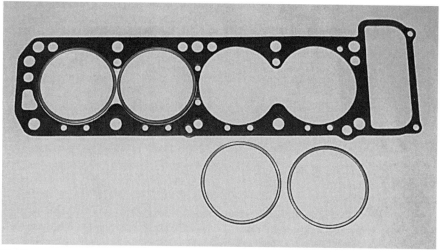

Sealing-ring head gasket for L-series four-cylinder engine is virtually bulletproof. Two of the rings are in position at left. Two rings at right will fit similarly into the gasket openings immediately above them.

or -12 (5/8- or 3/4-in. ID) union. Epoxy the fitting in place. You can now attach the dash- 10 or -12 pressure hose that goes from the pump to the fitting.

A dry-sump-pump's scavenging action creates negative crankcase pressure, eliminating the need for conventional crankcase venting. Plug the vent hole in the left front of the block with a 25mm core plug. Leave the cam-cover vent open or slightly restricted to produce the desired negative pressure—vacuum—in the crankcase. It should not exceed -5 in. HG below atmospheric pressure (vacuum) or the rear seal will "flutter" and leak.

To measure crankcase pressure during dyno testing, attach a hose from a water manometer to the dipstick tube. For a race setup, you can remove the dipstick tube. Tap the dipstick hole to 1/8-in. NPT and install a plug. Do this before you assemble the engine so metal cuttings won't get in the crankcase.

Cylinder Bores—Although always important, correct piston-to-bore clearance and bore finish are critical for a race engine. Just how much piston-to-bore clearance is needed depends on piston material and design. Therefore, this clearance is usually specified by the piston manufacturer.

Forged-aluminum racing pistons require more clearance than their cast counterparts. As a general rule, 0.004—0.006-in. more piston-to-bore clearance is needed, or 0.0055—0.0080-in. For a rebuild, up to 0.010-in. clearance may be acceptable.

Bore finish must be compatible with piston-ring material. Some engine builders deviate from the ring manufacturer's recommendations only because initial ring-seal break-in is more important than ring life. For example, they may leave a coarse bore finish to speed ring break-in time. Engine running time is relatively short between teardown and freshening up. However, most Nissan/Datsun engine builders use the hone finish recommended by the piston-ring manufacturer.

The crucial thing about a cylinder bore is its trueness. Bores must be as round and parallel—straight—as possible *after the cylinder head and crankshaft are installed*. This is because the torqued cylinder-head and main-bearing-cap bolts distort the block and, consequently, the bores. So, if the bores are true before the head and crankshaft are installed, they won't be afterward. Therefore, you must attempt to reproduce this condition during the boring and honing operations. Do so by installing the main-bearing caps and, if possible, a torque plate. Correctly torque the main-cap and cylinder-head bolts, page 86.

Final Cleaning—The final step in block preparation is cleaning. A race cylinder block must be absolutely clean with no remaining metal shavings or abrasive grit generated from the prep work.

Wash the block with clean solvent. Use nonmetallic bore brushes to clean the oil passages, threaded holes and cylinder bores. After brushing, thoroughly wash the cylinder bores with soap and water, then rinse clean. When the block is sparkling clean, inside and out, it must be completely dried to prevent it from rusting.

To reduce rust, liberally apply a water-dispersant, light oil, such as WD-40® or CRC®, to the entire block while drying with a compressed-air blow gun. Immediately afterward, wipe bores with a clean white-paper towel. The towel should come out perfectly clean—no dirt or particles of any kind. Don't forget to squirt oil into the oil passages and threaded holes. Then, apply engine oil to the cylinder bores and other machined surfaces. To protect the block, cover it with a large, clean plastic bag.

CHAPTER TWO
Crankshaft

Mike Rickman emerged the winner in SCCA's GT-3 class in 1983 at the wheel of his L20B-powered 200SX.

Fully counterweighted L20B crankshaft can be modified for racing. Crank can be substituted for Z20 crank that is not fully counterweighted.

All Nissan/Datsun OHC L-series engines have forged-steel crankshafts with cross-drilled main-bearing journals. Most of these crankshafts are suitable for racing applications.

There are a few crankshafts that are not desirable for racing or are easily replaced with a better one. The 1982–83 Maxima L24E crank has smaller connecting-rod journals than the early 240Z L24 crank. So, if you have the L24E Maxima engine you may want to use the 240Z crank and rods. The early L16 and late Z20 (Naps-Z) four-cylinder crankshafts are not fully counterweighted, while the later L16 and earlier L20B cranks are. Again, this is an easy substitution. The Z22 and Z24 four-cylinder crankshafts are also only partially counterweighted but there are no

Factory optional L28 crank has tapered counterweights, hollow crank pins and larger main-bearing-journal oil holes. This part is no longer available (NLA).

Factory-optional crankshaft (NLA)) has undercut radii on both main and connecting-rod journals. I don't recommend this modification because I've seen cracks develop in undercuts. Pressed-in oil-passage plugs should be replaced with 1/16-in. Allen-head pipe plugs.

Factory-optional crankshaft (NLA) has eight flywheel bolts. Black finish resulted from heattreating process.

factory-produced substitutes. Fortunately, the standard Z22 and Z24 crankshafts have both proved OK for racing.

Currently, the only two factory-optional race cranks available in the U.S. are for the L18 and L20B. There are optional cranks in Japan for the six-cylinder L24 and L28. Expense and the need to use special bearings are why these cranks are not offered in the U. S.

Except for the L28, aftermarket billet cranks are not popular for Nissan/Datsun engines. A billet crank is about three times more expensive than a race-prepped stock crank. The majority of billet cranks for the L28 are manufactured by Moldex.

CRANKSHAFT FEATURES

All Nissan/Datsun L-series engines have the same 49.97mm (1.967-in.) connecting-rod journals, except for the smaller 44.91mm (1.768-in.) rod journals of the '82-and-later L24E Maxima crank.

There are two main-bearing-journal diameters. The L16, L18, L24, L26 and L28 cranks have the small 54.95mm (2.163-in.) main journals; the L20B, Z20, Z22 and Z24 crankshafts have 5mm-larger 59.95mm (2.360-in.) main journals.

All standard and optional L-series crankshafts share the same type front snout, pulley-retaining bolt, keys, front seal, rear *register*—flywheel flange—rear seal and transmission input-shaft pilot bushing. Flywheel bolts are common except for the L18 optional crankshaft, which has larger bolts. Dampers, pulleys, crank sprockets and drive gears are interchangeable among all of the L-series engines.

L16 and standard L18 cranks have five 12mm flywheel bolts; L20B, Z22, Z24, L26 and L28 cranks have six. The optional L18 crank has six larger 14mm flywheel bolts and the factory-optional L20B and L28 cranks have eight 12mm flywheel bolts.

In addition to the standard cranks there are two other L-series cranks worth mentioning, the L13 and LD28. The L13 crank—installed in very early non-U.S. PL510s—can be used to destroke the L16 and L18.

The LD28 crankshaft—from the 2.8-liter diesel Maxima—can be used to stroke an L24, L26 and L28. These cranks share common features with other L-series cranks: bearing-journal size, front snout diameter and length, flywheel flange size and flywheel-mounting requirements.

Rod bolt tops just barley clear the bottoms of the cylinder bores. Check carefully here as some clearance grinding may be required. Stock bearings work with the LD28 crank.

CRANK PREPARATION

Mild Prep—Minimum crankshaft preparation is required for street-driven autocross and rally engines if power increase over stock does not exceed 50%. Essentially, all that is required is to be sure the crank is in like-new condition and is in correct balance.

To prepare a used crank, check the journals for size, roundness and finish. Also check straightness—*runout*—especially on a six-cylinder crankshaft, which is much longer than one for a

CRANKSHAFT SPECIFICATIONS
mm (in.)

Engine	Stroke	Main-Bearing Journal Diameter	Connecting-Rod Journal Diameter
L13	59.9 (2.358)	54.942–54.955 (2.16312.1636)	49.961–49.974 (1.96701.9675)
L14	66.0 (2.598)	54.942–54.955 (2.16312.1636)	49.961–49.974 (1.96701.9675)
L16	73.7 (2.902)	54.942–54.955 (2.16312.1636)	49.961–49.974 (1.96701.9675)
L18	78.0 (3.071)	54.942–54.955 (2.16312.1636)	49.961–49.974 (1.96701.9675)
L20B	86.0 (3.386)	59.942–59.955 (2.36002.3604)	49.961–49.974 (1.96701.9675)
L20A	69.7 (2.744)	54.942–54.955 (2.16312,1636)	49.961–49.974 (1.96701.9675)
L24	73.7 (2.902)	54.935–54.955 (2.16282.1636)	49.961–49.974 (1.96701.9675)
L26	79.0 (3.110)	54.935–54.955 (2.16282.1636)	49.961–49.974 (1.96701.9675)
L28	79.0 (3.110)	54.935–54.955 (2.16282.1636)	49.961–49.974 (1.96701.9675)
LD28 (Stroker Kit)	83.0 (3.268)	54.935–54.955 (2.16282.1636)	49.961–49.974 (1.96701.9675)
Z20	86.0 (3.386)	59.942–59.955 (2.36002.3604)	49.961–49.974 (1.96701.9675)
Z22	92.0 (3.622)	59.942–59.955 (2.36002.3604)	49.961–49.974 (1.96701.9675)
Z24	96.0 (3.779)	59.942–59.955 (2.36002.3604)	49.961–49.974 (1.96701.9675)

*1982-83 Maxima, 44.911 mm (1.768-in.) rod journal.

four-cylinder. Polish the journals carefully, then thoroughly clean the entire crank. Replace the transmission input-shaft pilot bushing.

Race Preparation—Selecting a crankshaft that's to be transformed into a race crank is the most important part of the preparation process. If you don't know the history of a used crank, inspect it closely for flaws and signs of past abuse.

The first step is to clean the crank thoroughly. Closely inspect for any damage. Inspect the condition of the journals and all other machined areas: snout, keyways, rear-main seal surface and threaded holes. Finally, check the crank for cracks by having it Magnafluxed. Although final crank-journal OD need not be checked until after you've heat-treated and polished the crank, double-check bearing-journal diameter against the above chart to ensure that journal sizes are to factory spec. Otherwise, the bearings may not give the correct oil clearance.

Oil-Gallery Plugs—If the crank passes all inspection steps, begin the actual preparation process. Remove the pressed-in oil-passage plugs. There's one in the end of each oil passage at each connecting-rod throw.

Chamfered oil holes in stock Nissan/Datsun cranks are good for racing as is or can have an entry flute added to the leading side of each hole.

Remove each oil-gallery plug by drilling a hole in its center. Then, using a conventional slide hammer with a sheet-metal-screw tip, pull out the plug. Once all plugs are removed, drill and tap the holes for 1/16-in. NPT Allen-head plugs. Don't install the plugs yet.

Polishing—Deburr and polish the entire crankshaft to remove surface stress risers that can initiate cracks. Deburring and polishing is easily done with a die grinder fitted with the appropriate stones and cartridge rolls.

Before you do any grinding or polishing, wrap the journals with several layers of masking or duct tape. This reduces the chance of scratching or gouging a bearing journal if you slip with the grinder.

The 1/16- or 1/8-in. NPT plugs that replace standard pressed-in plugs allow easier access to oil passages for cleaning.

Oil passages of L20B crank are modified by L.A. Billet Crankshaft Co. for engine builder Floyd Link. First four main journals are grooved. Oil passages are drilled so each of the first four main-bearing journals supply oil to one rod journal.

L20B crankshaft is carefully polished after heat-treating by L.A. Billet Crankshaft Co.

Use stones to get the rough stuff. Radius all machined edges and forging-line edges. Change to cartridge rolls for final smoothing and polishing.

Main-Journal Oil Grooves—An acceptable, but not common, practice is to groove the main journals that supply oil to the rod journals. Some engine builders even frown on the idea because of the fear of lost oil pressure. However, as long as there's sufficient oil flow, as with a dry-sump system, there should be no such concern. Regardless, the fear of crankshaft grooving remains even though some reputable Nissan/Datsun engine builders used this modification. For instance, Electramotive significantly reduced rod-bearing wear and fatigue by grooving the mains.

Grooving the mains is important for four-cylinder race engines. Otherwise, it's not needed. The oil supply for the number-two and -four main bearings supply all oil to the four connecting-rod journals. Therefore, only groove the number-2 and -4 journals of four-cylinder L-series engines. As for sixes, one main feeds one rod journal. So, oil starvation at the rod journals is not a problem under normal conditions.

Race-grade main bearings have grooves in both upper and lower shells. Any grooves put in the bearing journals should be sized and placed to match the bearing grooves. This will ensure that there's no reduction in bearing-surface area. Centered on each main-bearing journal, make the grooves 0.120-in. wide and 0.035-in. deep.

Heat-Treating—Heat-treating is a process that should be done after most of the prep work is done. Trying to deburr or groove the crank, or tap the oil holes for screw-in plugs after surface hardening would be foolish. Installing flywheel dowels in the rear register is the only machining operation that should be done *after* heat-treating. This is because high heat from the heat-treating process will distort the dowel holes.

There are several types of heat-treating processes available. Unfortunately, those processes that provide the best surface-hardness *penetration,* or *depth,* also increase distortion. Heat-treating processes that give minimal surface hardening also cause the least distortion. Regardless, *some distortion is always caused by heat-treating.* The most common types of crankshaft surface hardening by heat-treating are *Tuftriding, coolcase* and *nitriding.* Of these three, the 10-hour nitriding process seems to produce the best all-around results.

Most heat-treating processes require that the crankshaft be straightened afterward. This is done by striking a blunt-nose chisel that's butted against a bearing-journal radius. Scary! Straightening a crank by this method leaves marks on the radius as shown in photo at right. Although not detrimental to the crank, these marks can be disturbing if you discover them during inspection and don't know what they are and why and how they got there.

Polishing & Checking—All machined surfaces—bearing journals, snout and flywheel flange—require polishing after heat-treating. The smoother the main- and rod-bearing journals are, the better the bearings will wear. However, be careful. Excessive polishing will reduce journal size, resulting in excessive bearing clearance. And, excessive or careless polishing can also result in an hourglass-shaped journal. A journal that's smaller at the center will cause considerable trouble when you try to get the correct bearing clearance.

Unless you have total confidence in the work of your crank grinder/polisher, measure the journals extensively yourself. Measure each one in several positions along the journal's length for taper and around it for out-of-round.

After measuring each journal, check its finish. It may be desirable to hand-polish the journals yourself. Do this with 600- or 800-grit wet-or-dry sandpaper wrapped around the journal and taped in place. Now, wrap one full turn of 3/16- to 5/16-in.-diameter cotton rope around the sandpaper. Use the rope to rotate the sandpaper. Remember, use caution not to polish only the journal center or get carried away and polish too much. Periodically check your progress. Apply solvent frequently to the journal to speed the polishing.

Check Straightness—The final quality-control check to perform on your prepped crankshaft is to check its straightness. All you need for this is a dial indicator and two V-blocks.

Support the crankshaft on the V-blocks, one under each end main-bearing journal. Set up the dial indicator so its plunger is against and square to the center main-bearing journal, but out of line with the oil hole. If you don't have V-blocks, there's no problem. Just set the crank in an engine block that's turned

Crankshaft is checked for straightness on V-block fixture with dial indicator. Each main journal is checked for runout. Don't try to do this yourself, but crank is straightened by striking crank with blunt chisel and hammer.

Tell-tale marks in rod-journal radius resulted from straightening with blunt chisel as described. Raised material should be smoothed flush with surrounding surface.

Addition of three 3/8-in. dowels helps keep flywheel from shearing flywheel bolts.

Small-diameter, low-inertia flywheel was developed by Electramotive. Combination of flywheel and peg-drive double-disc clutch contributed to increasing safe rpm limit and life of L-series six-cylinder race engines. In 1999, the clutch has been reduced to a 5.5-inch diameter, but it is still mounted on a 10.5-inch flywheel.

Starter must be moved inboard when a small-diameter flywheel is used. This requires modifying the engine plate to clear Nissan/Datsun gear-reduction starter nose. Block must also be clearanced (arrow).

Electramotive crankshaft damper (both sides shown here) is available from Nissan Motorsports as 99996-E1060E. Special bolt/washer assembly E99996-E1065 should be used with this competition damper. Six-cylinder crank-fire-ignition paddle is custom-made and shown here only for reference.

upside down. Use only two main-bearing-shell halves in the block; the front and rear.

Rotate the crank to where the dial-indicator reading is lowest. Zero the indicator. Now, rotate the crank and note the highest reading; that figure is its runout. Double-check runout by repeating the process a few times. The crankshaft should have less than 0.001-in. bend, or runout.

Flywheel Dowels—Dowels that key the flywheel to the crank take some of the shear load that would otherwise be taken by the flywheel-mounting bolts. These loads consist of engine torque, crankshaft *torsionals* (torsional vibrations) and clutch inertia.

If you plan to use dowels to help keep the flywheel on the crankshaft, now is the time to install them. Use three 3/8-in.-diameter, 7/8-in.-long dowels on the same circle as the flywheel-bolt circle.

To install flywheel dowels, you must drill holes both in the crankshaft and flywheel. However, before you get out your 1/2-in. electric drill and 3/8-in. drill bit, stop! Drilling a crankshaft and flywheel for dowels is a machine-shop operation. As with any engine-machining operation, accuracy is a must.

The machine shop will need a drill jig to locate the drill accurately for drilling the flywheel-dowel holes. Using a U-letter drill, drill three evenly spaced holes 0.500-in. deep in the rear face of the crankshaft register on the flywheel-bolt circle. The evenly spaced holes ensure that crankshaft balance is not affected by the dowel installation.

After drilling, ream the holes to 0.3740—0.3745 in. The 0.0005—0.0010-in. interference fit will give the 3/8-in. diameter, 7/8-in.-long dowels the correct press fit. After reaming, coat the dowels with red Loctite® and press them in. They should be installed so they are 0.030—0.050-in. below the flywheel face. Measure the flywheel-hub thickness to determine how far the dowels should project from the flywheel flange. The dowel holes in the flywheel are also drilled with a U-letter drill using the drill jig, but reamed to 0.375 in. to give a zero fit to the dowels.

Pilot Bushing—Install a new pilot bushing or needle bearing in the crank's rear register. A needle roller bearing is available from Nissan Motorsports as 32202-09500. However, the standard, stock bushing is acceptable for racing applications.

Bearing Selection—To select bearings to achieve correct clearances, measure and record journal diameters. Measure each journal in five or six places. These dimensions will be needed later during engine assembly.

NISSAN MOTORSPORTS CRANKSHAFTS

The 2.8 Big Bore Stroker Kit allows you to increase displacement to 3098cc (193ci). Kit includes LD28 crankshaft, L24 connecting rods, 89mm pistons and rings. Pistons have a positive deck height of approximately 0.025 and must be machined to accommodate various block and/or cylinder-head combinations. Compression ratio can be adjusted by using either of the two 91mm cylinder-head gaskets.

Special racing cranks for the L18 and L20B are available from Nissan Motorsports. Although now discontinued, factory competition cranks for the L16 were available, so you may be able to find one.

Prototype damper on L28 Turbo reduced crankshaft torsional deflection (torsional vibration). This damper was never produced commercially.

L16, L18 and L20B competition cranks are fully counterweighted, properly heat-treated and have *round* journals. The journals are perfectly round because they were ground *after* heat-treating. This method gives more accurate bearing journals than heat-treating the finished crank, then polishing it.

Nissan Motorsports crank-bearing journals also have larger fillet radii. And, those for four-cylinder engines have different oil passages. Also available are optional four-cylinder race cranks drilled so each main-bearing journal supplies one rod journal with oil rather than supplying two rod journals.

AFTERMARKET BILLET CRANKS

Billet cranks for L28 six-cylinder engines are popular. For instance, Bob Sharp Racing has successfully used Moldex billet cranks in their engines for many years. Because these cranks are heat-treated before grinding, the finished journals are, like Nissan Motorsports cranks, truly round. And because these cranks are custom manufactured, the counterweights, journal sizes, stroke and oil flow can be altered as required.

L.A. Billet has manufactured some Nissan/Datsun L-series cranks for use in IMSA racing. These billet cranks, which use Cosworth-size rod journals, are available with several different strokes.

If you are considering a billet crank, be prepared for a wallet shock. They cost two to three times the price of a race-prepped stock crank.

CHAPTER THREE
Pistons & Rings

Don Devendorf won 1982 IMSA GTO Championship in his "awesome" 280ZX Turbo. Car was so competitive it out-qualified many GTP cars. Mark Yeager photo.

Nissan does not offer "hot-rod" replacement cast pistons, but a wide variety of pistons from L-series engines are interchangeable. This piston interchangeability allows for many practical bore-size and compression-ratio combinations.

However, cast pistons, regardless of manufacturer, have limitations. For instance, load limits can be exceeded with a high compression ratio or under turbocharger boost. The detonation that may result, especially from today's low-octane pump gas, will destroy a cast piston.

The limitations of a cast Nissan piston are hard to estimate, but it's safe to assume that the average non-high-compression street-car engine will survive with cast pistons. An exception may be a turbocharged engine with boost pressures exceeding 12 psi.

By using the accompanying piston comparison chart or Nissan/Datsun service manuals, you can probably find a suitable cast piston for most applications. An important consideration when using pistons not meant for your engine is to use the ring set specified for the pistons, not those listed for the engine. This is important because the width of the rings must match the width of the grooves in the piston.

STOCK NISSAN PISTON RINGS

Standard Nissan/Datsun piston rings are ideal for most street applications. As discussed in Chapter 1, bore finish must match ring-face material, regardless of the type of ring used. A competent engine machinist will know what type of hone finish to use, providing the piston and-ring set accompanies the block. With this in mind, send all three items—pistons, rings and block—to the machine shop when having a

Pistons & Rings

Although stock Nissan/Datsun cast pistons fit 85mm-bore L18 or L20B, tops are significantly different. Piston at left is optional SSS L18 flat-top piston, P/N 12010-A8720. Stock L20B piston, center, has dish of 11.36cc to reduce compression. LZ20E piston at right has 0.250-in. less pin height.

Side view of pistons shown at left: Two at left have same pin height-distance between pin-bore center line and top. LZ20E piston at right with 6.4mm (0.25-in.) less pin height would work well in street turbocharged L20B.

STOCK NISSAN CAST PISTONS

Application	Nissan Part Number	Bore Size in./mm	Pin Height in./mm	Piston Dish Volume (cc)	Nissan Ring-Set Part Number
L16	12010-23002	3.268/83	1.50/38.1	7.01	12033-23000
L16 + 1mm	12010-22006	3.307/84	1.50/38.1	7.01	12038-A3500
L24	12010-E3111	3.268/83	1.50/38.1	0	12033-E3100
L24 + 1mm	12010-E3116	3.307/84	1.50/38.1	0	12038-E3116
L24	12010-H2711	3.268/83	1.50/38.1	0	12033-E3100
L18	12010-U2001	3.346/85	1.50/38.1	4.36	12033-A8702
L18 + 1mm	12010-A8706	3.386/86	1.50/38.1	4.36	12038-A8702
L18	12010-A8720	3.346/85	1.50/38.1	0	12033-A8775
L20B	12010-U6001	3.346/85	1.50/38.1	11.36	12010-U6001
L20B + 1 mm	12010-U6006	3.386/86	1.50/38.1	11.36	12038-U6001
LZ20S up to 7-84*	12010-W4001	3.346/85	1.40/35.56	N/A	12033-N8502
LZ20S + 1 mm*	12010-W4003	3.386/86	1.40/35.56	N/A	12038-N8520
LZ20E**	12010-N8711	3.346/85	1.25/31.75	N/A	12033-N8520
LZ20E + 1mm**	12010-N8713	3.386/86	1.25/31.75	N/A	12038-N8520
L28	12010-Y4111	3.386/86	1.50/38.1	10.9	12033-N4200
L28 + 1mm	12010-Y4116	3.425/87	1.50/38.1	10.9	12038-N4200
L28 Turbo	12010-P9012	3.386/86	1.50/38.1	10.9	12033-P9010
L28 Turbo + 1mm	12010-P9017	3.425/87	1.50/38.1	10.9	12038-P9010
L28 (After 7-80)	12010-P7912	3.386/86	1.50/38.1	0	12033-P7910
L28 + 1 mm (After 7-80)	12010-P7917	3.425/87	1.50/38.1	0	12038-P7910
LZ22S*	12010-06W11	3.425/87	1.38/35.0	9.32	12033-06W20
LZ22S + 1mm*	12010-06W13	3.465/88	1.38/35.0	9.32	12038-06W20
LZ22E (From 1-82)**	12010-D8101	3.425/87	1.28/32.5	9.32	12033-06W20
LZ22E + 1mm (From 1-82)**	12010-D8103	3.465/88	1.28/32.5	9.32	12038-06W20

*LZ20S or LZ22S = Naps-Z engine with carburetor
**LZ20E or LZ22E = Naps-Z engine with electronic fuel injection

High-compression, forged Venolia piston fits 87mm (3.425-in.) 1mm overbore L28. Floating pin is retained with Teflon buttons. Piston is available from Nissan Motorsports.

Venolia L20B piston with finish-machined valve pockets and dome. Pin size is 0.866 inch. First and second ring grooves are 1mm wide.

This type piston-dome configuration must be custom-ordered or finished by engine builder. This piston was for Don Devendorf's 2.5-liter GTU engine.

Piston domes that are custom fitted to cylinder-head combustion chambers can be complicated. This dome fits early E3100 cylinder head for SCCA GT-2 engine.

block bored and/or honed. Also, if you want a different piston-to-bore clearance than that recommended by the service manual, specify the desired clearance.

FORGED RACING PISTONS

Forged high-compression *domed* racing pistons are readily available for most L-series engines. Nissan Motorsports stocks a large selection of forged pistons for most popular applications. Cosworth, Arias, Ross and Venolia manufacture forged pistons for Nissan engines.

Cosworth is probably the only piston manufacturer that will not produce a custom piston for your application. This is unfortunate because of their high quality. However, the domes of some Cosworth pistons can be modified to fit many combustion-chamber configurations.

Forged racing pistons are available off the shelf from Nissan Motorsports in only one configuration for each engine. However, these pistons can be modified to fit many applications. Like Cosworth pistons, Nissan Motorsports pistons have domes that can be machined to fit a specific combustion chamber. Machining the piston dome is preferred over grinding the combustion chamber to fit the piston—more

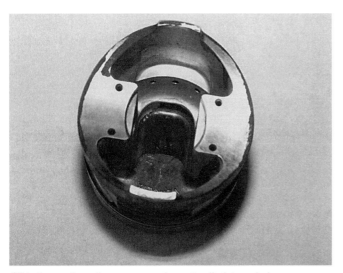
Oil holes such as these are good way to oil piston-pin bore.

Arias piston for L16 has single oil hole drilled horizontally in and intersected by diagonal hole that runs down to pin bore. Note Teflon pin button.

about this later. Most forged pistons have enough dome thickness to allow some machining in the valve-pocket area for additional piston-to-valve clearance.

If you can tolerate a few weeks- to months-wait, order pistons to your specifications from a piston manufacturer such as Arias, Ross, Venolia or J.E.. They produce high-quality forged pistons in virtually any configuration. Bore size, ring types, pin height, pin diameter, pin-oiling method, valve-pocket size and depth, and dome can all be machined to your requirements. This is an advisable consideration if you need a non-stock pin size, bore size or valve-pocket depth. Be sure to provide an accurate drawing with all details spelled out when ordering custom pistons.

I've found it beneficial to fit a piston dome to the smallest possible cylinder-head combustion chamber rather than opening up the chamber to fit an existing piston. To obtain maximum compression and still have adequate piston-to-valve clearance, order pistons with the correct-valve-pocket depth. If the valve pockets are too deep, compression will decrease.

Also, if the pockets require much additional machining, piston-dome thickness at the base of the pockets may be inadequate. Dome thickness at any point should be 0.200 in. minimum.

A standard 21mm (0.827-in.) diameter piston pin is not adequate unless it is made of very high-quality material. Consequently, it is best to use a 22mm (0.866-in.) diameter pin because its larger size is stronger and gives more load-bearing area.

Pin Oiling—Pin-oiling holes in the piston-pin bosses provide lubrication to the pins and their bores. Arrangement of these oiling holes varies. Some are good, some are bad. For structural reasons, avoid those that intersect the vertical centerline of the pin from the top or bottom of the pin boss. Such oiling holes create stress risers at the highest-loaded, thinnest sections of the pin bosses. Thus, the potential for failure is highest.

A desirable pin-oiling-hole arrangement has two holes from the underside of each pin boss, one on each side that intersects and is tangent to the pin hole. A single oil hole that connects the top of the pin bore with the oil-ring groove should be at an angle to the vertical plane of the pin-bore centerline. To repeat, under no conditions should there be vertical oil holes on the pin-bore centerline in the top and bottom of the pin bosses.

Street Turbocharged Engines—Turbocharging the L-series engine is popular with many people, including Nissan. However, not many turbocharged L-series engines are found in competition. Regardless, those that have been used in racing have been successful. Don Devendorf/Electramotive and Bob Sharp Racing are almost synonymous with turbocharged Nissan racing engines. However, in spite of the success of the few, street turbocharging is where most of the activity is.

For street-turbo applications, low-compression forged pistons with the top compression ring moved down on the piston and away from combustion heat are popular. Most of these pistons have a thick, flat top that has been lowered to reduce compression. These *turbo* forged pistons have significant strength advantages over stock cast-aluminum pistons.

PISTON PINS

The subject of piston pins and how they are retained in their pistons can generate numerous opinions. These opinions range from which type of pin to use, such as thin-wall, lightweight pins, versus heavy, strong pins, to how to retain the pins, to full-floating versus press-fit pins.

Street Performance—As I've already stated, for the majority of street applications, stock cast-aluminum Nissan pistons are acceptable. These pistons come with good-quality pins that are more than adequate for street use. Also, the pins are pressed in the connecting rod. So, for street use, retain the stock pins and the pressed-fit setup.

When installing a press-fit pin into its piston-and-rod assembly, the operation starts with inserting the pin into its bore at one side of the piston. The pin is then pressed through small end of the rod and into the other side of the piston until it's centered in the piston-and-rod assembly. The problem with assembling a rod and piston in this manner is that damage can result. Of major concern is the chance of cracking or breaking the piston bosses at the pin bore. This can result from not properly supporting the piston while pressing in the pin.

To eliminate the chance of cracking the piston-pin boss during assembly, some automotive machine shops heat the small end of the rod red hot to expand it. This also reduces the force required to press the pin through the connecting rod and lessens the chance of damaging the piston.

If aftermarket piston/pin assemblies are used, measure the pin diameter and compare it to the 0.8265—0.8267-in. (20.993—20.998mm) diameter of the standard Nissan/Datsun pin. Any difference in pin diameter will change pin interference in the small end of the connecting rod. The pin should be 0.0006—0.0014-in. (0.015—0.035mm) larger than the small end of the rod for a correct press fit.

Whatever method is used to retain the pin, it should be centered in the rod and piston. The piston must rotate freely on the pin. This ensures that the pin won't gall the piston-pin bore and the piston will rock with minimum resistance as it reciprocates while traveling up and down the cylinder bore. Cast-aluminum pin-to-piston bore clearance is 0.0002—0.0005 in. (0.006—0.013mm).

Racing—For racing, the piston pin should be *free-floating*—there's clearance to the pin in the small end of the rod, not interference. Piston- and rod-to-pin clearance should be 0.0008—0.0011-in. (0.020—0.028mm). This applies to most forged pistons and connecting-rod small ends with either a honed-steel finish or thin-wall bronze bushings.

Aftermarket forged racing pistons are usually sold with pins. This means the pins are fitted to the pistons. So, they should have the correct clearance in the piston.

The material a pin is made from is as important as the pin clearance. Some racing-piston manufacturers use extruded steel tubing. This material has serious flaws on the surface of the ID, which can lead to pin breakage.

Other manufacturers use thin-wall, lightweight piston pins that may eventually fatigue and, in some instances, break. This may occur even though thin-wall piston pins are usually quality-crafted with excellent material.

The ultimate piston pin should have a larger outside diameter, or 0.866 in. (22mm) versus the 0.826-in. (21mm) stock diameter. The pin should be machined from solid bar stock 8620 steel. This material is carefully bored and honed to a relatively thick-walled, hollow pin. The pin is then polished and hard-chrome plated.

This ultimate straight-bore full-floating piston pin can be retained with Teflon buttons, one at each end of the pin. Nissan-Motorsports forged racing pistons come equipped with heavy-wall, straight-bore, standard-diam-

Forged piston for 3.0-liter turbocharged L28 street engine. Piston and late L28 Turbo cylinder head reduce compression to a manageable figure.

Black on piston top was caused by oil consumption due to poor ring seal. Experimental 2.5mm (0.100-in.) wide oil ring didn't provide sufficient sealing.

The two disadvantages of forged pistons for the street are higher cost and additional piston-to-bore clearance. Higher cost is self-explanatory. Most everyone understands the limitations of his pocketbook. Extra piston-to-bore clearance also causes *piston slap*—undesirable on a street engine. When the engine is cold and the pistons haven't warmed up and expanded to fit their bores, they are noisy. Also, if the engine is operated at wide-open throttle when cold, the extra *wobble* of the pistons in their bores can cause skirt and ring damage.

RACING PISTON RINGS

Ring Material—The question of which piston-ring material to use for

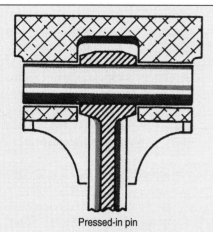
Pressed-in pin

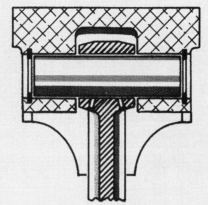
Full-floating pin with retainers

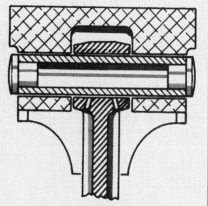
Full-floating pin with pin buttons

eter pins with Teflon buttons in the ends. The buttons install in counterbores in the pin ends. These pistons are designed with top and bottom support for the buttons in the skirt. Additionally, the center of the buttons index into the center of the pin bore for additional support. This type of pin-and-button setup has been very successful for customers of Nissan Motorsports and those with Electramotive-built engines.

Aftermarket Nissan racing pistons are also available with other types of pin-retention methods. The most popular are Spirolox® internal-type retaining rings installed in each end of the full-floating pin bore. While Spirolox are excellent, they are hard to put in and take out. Some engine builders don't trust circlips for the high-RPM engines.

The piston must have the correct-size groove—width and depth—cut in each end of the piston-pin bore to accept the particular retaining ring. Some engine builders utilize two retaining rings at each end in grooves sized accordingly.

Regardless of whether one or two rings are used at each end, the grooves must be positioned in relationship to the pin so no preload is created on the installed clips, or have so much clearance that the pin can move excessively, thus generating excessive thrust load against the clips. Ideal pin-to-retainer clearance is *zero*.

If you're wondering which way to go, here's what I recommend: I prefer Teflon buttons with full-floating piston pins. Why? They are nearly "bulletproof." They are also easy to assemble and service. The only crucial requirement for a Teflon-button-retained pin is that the total pin/button assembly has the correct end play. Pin end play is determined by end clearance to the bore wall as it's installed in the piston. This clearance should be 0.015—0.017 in. (0.38—0.43mm) for most L-series engines.

Connecting-Rod Small End—For sizing or preparing the small ends of the connecting rods for pressed-in pins or floating pins, is discussed in Chapter 4.

racing is as controversial as which is the best ring design for racing. However, it is generally agreed that plain cast-iron or cast-steel rings work well on road- or dragrace engines where quick break-in is required.

Chrome-face rings are best for offroad racing or rallying where the hard ring face is more wear-resistant to dirt. Use moly-faced rings similar to those used in original-equipment Nissan engines for a multi-purpose race engine or one that's to be run for extended periods in durability racing.

Ring Design—Regarding ring design, one point that is generally agreed upon is that narrow—1mm-wide—compression rings are advantageous in engines that turn over 7500 rpm. Total Seal® gapless-type rings have been used in the second ring groove with some success as have *stepped-scraper* rings. Ring sets for road racing usually have a square top ring and stepped scraper-type second ring. Oil rings are usually the three- or four-piece *medium-tension* type for minimum drag.

The rings you would like to use may not be available in +0.020 and +0.040 over L28 sizes. Work with the piston manufacturer to get the best-available rings.

PISTON & RING FIT
Piston and ring fit are crucial for durability and bore sealing. Ring, piston and bore diameters must all match. And the rings must fit the piston grooves. Finally, cylinder-bore finish must be compatible with ring material.

Clearances—Each piston manufacturer provides a piston-to-bore-clearance specification for each of their pistons. Also specified by the manufacturer is where to measure piston diameter. Pistons are ground on a taper, resulting in a smaller diameter at the top than that measured across the bottom or at the skirt tip.

Why are pistons tapered? Taper is needed because a piston's temperature differential from its top to its bottom results in different expansion rates. Specifically, the top of a piston is directly exposed to combustion-chamber heat, resulting in considerable expansion. Heat from the piston top travels down the ring belt and to the ring belt and to the skirts. As

Wide selection of ring sets covers virtually all bore-size L-series engines.

heat is conducted down the piston, much of the heat that would otherwise go to the skirts is transferred to the cylinder walls and oil. Thus, piston skirts operate much cooler and expand much less than the top. Consequently, more piston-to-bore clearance is needed at the top than at the bottom.

As mentioned previously, piston-to-bore clearance is greater for forged pistons than for cast pistons. However, within each group, clearance and taper varies significantly from one aluminum alloy and piston design to another. Therefore, always follow the piston manufacturer's recommendations when building your engine.

Clearances—Regardless of piston design, piston-to-cylinder-head clearance-not to be confused with piston-to-valve clearance-should never be less than 0.050—0.055 in. on high-speed engines. Piston-to-pin clearance on forged pistons should be 0.0008—0.0011 in. and 0.0002—0.0005 in. for cast pistons.

The fit of a ring in its groove is crucial to proper ring performance. For example, excessive ring-to-groove side clearance will allow ring flutter at high engine speed, resulting in broken rings or poor ring-to-bore sealing. Side clearance for a cast piston should be 0.0016—0.0029 in. on the top compression ring and 0.0012—0.0028 in. on the second one. With forged pistons, use 0.0025—0.0035-in. side clearance on the top ring and 0.002—0.003 in. on the second ring. Backside clearance on the first and second rings should be 0.000—0.020 in. for both forged and cast pistons.

Ring end gap on the first and second rings is also important. Excess gap allows combustion leakage. Insufficient end gap that allows the ring ends to butt, will cause ring breakage.

A good rule to follow for top compression-ring end-gap clearance is 0.005 in. for each inch of bore diameter. Use a little less for the second ring, or 0.004-in. end clearance per inch of bore diameter. Using this method to determine end gaps, most L-series race engines should have 0.017-in. end gap on the top ring and 0.014-in. end gap on the second ring. Here is how these figures were arrived at:

With a 3.4-in. bore, end gap is 3.4 in. X 0. 005 in./in. = 0. 017 in. for the top ring. End gap for the second ring is 3.4 in. X 0.004 in./in. = 0.014 in. Using this method for determining end gap assumes that the cylinder bore is round and has little or no taper. A bore with 0.001-in. taper will give about a 0.003-in. variation in end gap as the piston and rings move up and down in the cylinder.

Oil-ring-rail end gap is not critical. This is evidenced by the wide allowable specification: 0.010—0.040-in. end gap is OK.

CHAPTER FOUR
Connecting Rods

Datsun parade: Paul Newman leads Jim Fitzgerald and Bob Leitzinger during running of 1980 SCCA National Championships at Road Atlanta. Race-prepped stock L-series rods are more than adequate for such applications.

All stock L-series Nissan/Datsun connecting rods are forged steel. Pressed pins are used at the small ends and 8mm or 9mm bolts retain the bearing caps at the big ends. Except for the '82 and '83 Maxima L24 rod, big- and small-end diameters are the same at 53mm (2.087 in.) and 20.97mm (0.825 in.), respectively. The L24 Maxima rod is 5mm (0.197 in.) smaller at the big end and 1mm (0.039 in.) smaller at the small end.

Except for the Maxima rod, stock L-series connecting rods are OK for street-performance and most race applications. However, for use in race engines, stock rods require some preparation to ensure they will endure the rigors of higher rpm and power output.

MILD PREP
Street, hot-rod and mild-competition applications generally only require that the stock rods be in excellent condition. They must be free of flaws, accurately machined, balanced as a set, and fitted with *new* bolts.

Selection—The preparation of stock connecting rods begins with selecting four or six rods that can be correctly balanced. This is important, particularly for a race engine. Nissan/Datsun rods don't have balancing pads. Consequently, removing large amounts of weight—material—can adversely affect strength. Removing material is also time-consuming.

Choosing a set of rods from a factory-assembled street engine is a good way to ensure you're getting a good set. They should be, at the worst, close to being in balance. However, if you put together a set from rods obtained separately, such as from the parts counter, it's possible that they

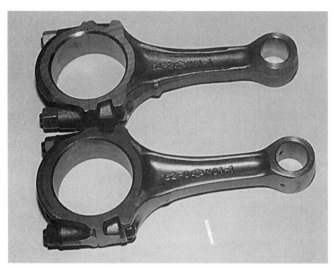

Stock L16/L24 connecting rod is at top; race-prepped version of same rod is below.

Big ends of rods are sized, or honed to 2.0866-2.0871 in. (53-53.01mm), to ensure correct bearing crush. Tom Monroe photo.

STOCK NISSAN CONNECTING RODS

Application	Nissan Part Number	Center-to-Center Length (mm/in.)	Bolt Size	Notes
L16 4-72 & later	12100-N0102	133mm/5.240	9mm	Preferred for racing
L24 5-71 up to 6-81	12100-N0102	133mm/5.240	9mm	Preferred for racing
L24 7-81 & later	12100-D0210	133mm/5.240	8mm*	
L24 7-81 & later	12100-D0220	133mm/5.240	8mm*	
L16 up to 7-70	12100-23000	133mm/5.240	8mm	
L24 up to 3-71	12100-23000	133mm/5.240	8mm	
L16 8-70-7-71	12100-N0100	133mm/5.240	8mm	
L16 8-71-3-72	12100-N0101	133mm/5.240	8mm	
L24 up to 4-71	12100-E3002	133mm/5.240	8mm	
L18 all	12100-A8703	130.2mm/5.132	9mm	
L26 all	12100-A8703	130.2mm/5.132	9mm	
L28 all	12100-A8703	130.2mm/5.132	9mm	
L20B all	12100-U6000	145.9mm/5.750	9mm	Preferred for racing
LZ22S '720	12100-U6000	145.9mm/5.750	9mm	Preferred for racing
LZ22E 200SX	12100-D8110	145.9mm/5.750	9mm	
LZ20 all	12100-N8500	152.45mm/6.00	9mm	

*Effective 7-81, L24E connecting rod was redesigned. Piston-pin-bore diameter was reduced to 19.97mm (0.786 in.). Big-end-bore diameter was reduced to 47.95mm (1.888 in.). Bolt size was reduced from 9mm to 8mm. Center-to-center length was retained.

can't be balanced correctly.

Magnafluxing—Replace *all* of the bolts with new ones, then have the rods *and* bolts Magnafluxed. If any flaws are found, replace the rod or bolt. Once you are assured that the rods and bolts are free of flaws, check alignment—straightness—of the rods. You could also check *center-to-center length*—distance between big- and small-end centers-but it's not necessary because this dimension is "always" correct.

The size and roundness of the big end is not as consistent as center-to-center length. Therefore, each should be *sized*, or honed to the correct size. When you deliver the rods for sizing, supply new bolts with them. For the sizing operation, the rods must be fitted with the bolts they'll be used with.

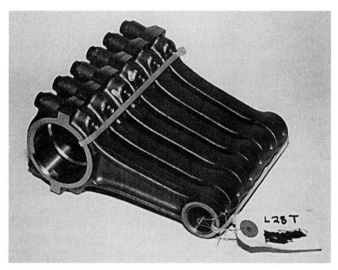
Six Nissan/Datsun sports-option rods were selected from inventory and wrapped as a set suitable for final balancing. Rods were eventually used in Don Devendorf's IMSA GTO 280ZX Turbo. These rods are no longer available.

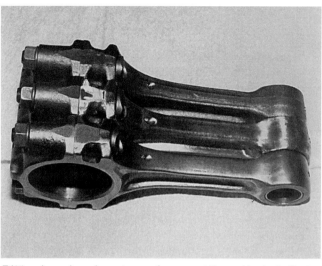
FJ20 rods are in various stages of race prep. Top rod is stock. Center rod is ground and polished. Bottom rod is polished and shot-peened. When finished, 5.51-in. (140mm) long rods were installed in IMSA GTU L28.

Sizing Rods—The rod and cap parting surfaces are ground on a special grinder to reduce the top-to-bottom big-end diameter. Then the new bolts are installed. The cap can now be fitted to the rod and the rod-bolt nuts torqued to the *lower limit* of the torque specification: 23 ft-lb for 8mm bolts; 33 ft-lb for 9mm bolts. Use the manufacturer's specification if special bolts are used. Sometimes, it may be necessary to *pinch*—bend together—the big end to reduce the diameter measured across the parting surface. This reduces the big-end diameter full circle.

The big ends of the rods are now ready for honing. Hone them to a diameter of 2.0866-2.0871 in. After the rods are sized, measure the big ends yourself to confirm they are correct.

The big-end bore diameter affects *crush* on the rod bearing inserts. It also affects the resulting bearing clearance.

Although you shouldn't need to do any work at the small ends, check them. Their diameters should measure 0.0006—0.0013 in. *smaller* than the pins to give the correct press fit.

Balancing—The next step is to balance the rods as a set. Balancing assures that all rods and their big and small ends weigh the same. The small end is part of the engine's reciprocating mass and the big end is part of the rotating mass.

When finished, the small end of the rods should weigh within 1.0—1.5 grams of each other; the big end of the rods should weigh within 1.5—2.0 grams of each other. These tolerances are suitable for moderate-speed engines. The typical L16 connecting rod—5.24-in. (133mm) center-to-center length—weighs about 680 grams ±60 grams.

Weight to balance the small ean can also be removed from inside of the piston pin.

RACE PREP-STOCK RODS
Selection—Race-prepping stock rods begins the same as for a mild prep; by selecting them. The reason is the same, too: To balance a set of stock connecting rods accurately without removing large amounts of material from one or more of them depends on the set you begin with.
Rough Balancing—Once you've selected the rods, have them rough-bal-

anced. This ensures that final balancing, which is done after most of the prep work is finished, can be done with minimal material removal.

Polishing—This is where preparing stock race rods differs from mild prepping. After Magnafluxing, polish the rods. This time-consuming process will reduce the possibility of cracks starting as the result of surface irregularities. You'll need your die grinder and some rotary stones and cartridge rolls.

Start by grinding and polishing smooth the forging line on both sides of the beam. Also, radius the beam edges. Move the grinder along the length of the rod. Keep grinding marks parallel to the long axis of the rod. The *web*—section joining the flanges at the center of the beam—generally does not require grinding or polishing.

The edges of the bolt-head recess and inside corners must be smoothed. Give the cap the same treatment. All sharp machined edges and forging edges should be smooth when you're finished.
Pin Oiling—If you plan to convert the piston pins to the full-floating type—as opposed to the stock-type pressed pin—

36 How to Modify Your NISSAN/DATSUN OHC Engine

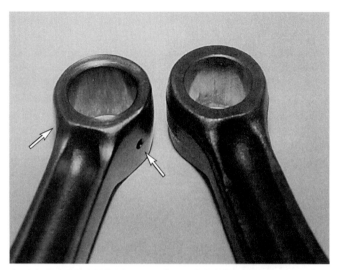

At right is stock L16/L24 connecting rod. At left is prepped rod. Note pin-hole location (arrows). A similar hole (not visible) is on the opposite side of the pin boss in the same location.

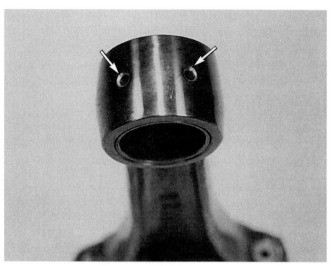

L28 Nissan sports-option connecting rod shows factory-preferred pin-bore oiling holes (arrows). If pin bore appears to be large, it is. Pin bore measures 23.5mm (0.927 in.) versus stock 21 mm (0.826 in.) pin bore. This rod is no longer available.

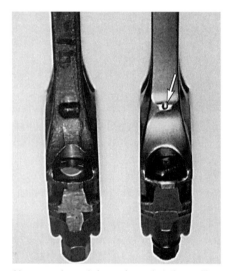

Note punch mark (arrow) used to close oiling hole of race-prepped rod. Prepped rod uses 3/8-in. bolts versus 9mm stock rod bolts. Detailing to area around bolt recess is crucial to reduce fatigue in highly stressed area.

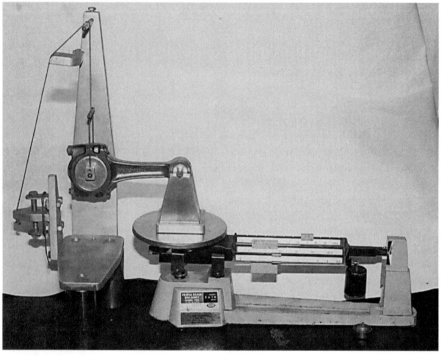

Balancing connecting rods with fixture and scale. Setup and consistency are crucial to maintain accuracy. Small end is being weighed.

the pin bore will require lubrication.

Oil-supply holes can be drilled in one of several locations. The most popular small-end oil-supply hole is at the top center of the rod. Make this hole about 1/8 in. (3mm). Another possibility is to drill a pair of smaller holes—about 3/32 in. (2.25mm)—on the underside of the pin bore at each side. Electramotive drilled a pair of 3/32-in. (2.25mm) holes, one in each side at about a 45 angle to the beam centerline.

After drilling, countersink the holes. This will break the sharp, rough edges and funnel additional oil to the holes.

For how to convert to floating pins, turn to page 38.

Skirt Oil Hole—The additional rod-bearing clearance used in a race-prepared engine and the relatively high oil pressure used will result in too much oil being sprayed on the cylinder walls/pistons. Therefore, if the rods each have an oil-squirt hole, close it off.

You'll find the skirt oil hole immediately above the big end on the *thrust side*—right side when installed in engine—of the beam.

To close the squirt hole, peen it shut. Use a center punch and hammer to do this. Once the hole is closed, polish the area thoroughly.

Final Balancing—You can now final-balance the rods. Afterward, the small ends of the rods should weigh within 1.0-1.5 grams of each other. The same goes for the big ends; weight should be within 1.5-2.0 grams of each other.

Shot-Peening—After balancing, have the rods *shot-peened*. To keep the big and small-end bores from being shot-peened, mask them with tape. Also, install the caps with an old set of modified bolts and nuts. To modify the bolts, grind down the bolt heads so the shot will have access to the radius at the corner of each bolt-head flat. This crucial radius should receive maximum shot-peening.

Next, you'll have to find someone to do your shot-peening. The best place to start is in the Yellow Pages of your phone book under HEAT TREATING. If the people at the other end of the line can't steer you in the right direction, call an aircraft-engine repair shop. Next, try a machine shop. You'll find that most shot-peening facilities specialize in aircraft parts and similar critical components. Consequently, they should know which precise shot-peening process to apply to an auto-engine connecting rod.

Rod Bolts—The stock rod bolt is the weakest link in the Nissan/Datsun L-series rod. An easy way to remedy this problem is to replace the stock rod bolts with 3/8-in. bolts from the 396-CID big-block Chevy passenger-car engine. Or, you can go one better. A variety of high-performance aftermarket bolts can be used as replacements. If you use the Chevy bolts—standard production or aftermarket—the installation operation is *not* standard. And, it can't be performed at the corner machine shop.

SHOT-PEENING

Most shot-peening facilities will know the requirements for your work, but you should be informed so you can ask intelligent questions concerning any operation you request. Let's start with a definition of shot-peening.

Shot-peening is a surface-treatment process for strengthening metal against *cohesive* failures by *prestressing*—putting a stress into a part before it is loaded. Cohesive failures are caused by *stress risers*—such as a chip in a windshield—stress corrosion, or fatigue from tension-induced overstress.

Shot-peening is the most versatile method of prestressing metal surfaces. It can be applied to any shape at relatively low cost. And, when properly performed, a 50% increase in *yield strength* can be expected. Yield strength is the stress a part can withstand immediately before it deforms, or stretches.

Although you probably won't be involved in the process, here's how it works. Cast-iron or steel shot—similar to that from a shotgun shell—is directed at high velocity against a part to create indentations in its surface. The indented surface tries to expand and occupy more space. But, because it can't, the surface is forced into *compression*, thus prestressing the part.

Shot-peening quality is determined by the depth of the prestress. Prestress depth is, in turn, a function of the hardness and ductility of the metal, the state of, *strain* of the metal—loaded condition—when peened, and the characteristics of the shot and shot stream. The degree of coverage, or number of shot impacts for each square inch of area, shot size, direction relative to the work and velocity, and the condition of the peening shot also contribute to shot-peening effectiveness.

Shot-size selection depends on the section thickness of the metal part, surface indentation desired, surface irregularities and fillet radii.

Shot velocity depends on the mass of the shot, shot size, shot hardness, and the direction of the shot stream relative to the metal part.

Shot cleanliness is required to achieve good peening. Reasonably round, uniform-size shot that's free of small fragments and dirt must be used. Otherwise, the shot will cut rather than peen, contaminate, and reduce peening effectiveness.

Shot quality refers to the size, shape roundness, hardness and impact fatigue strength of the peening shot.

Coverage refers to distribution of the shot-peening stress on the surface of the work. Uniform stress means that the impact points are so numerous that the induced stress of each impact overlaps the adjacent stressed area to produce a uniformly prestressed part.

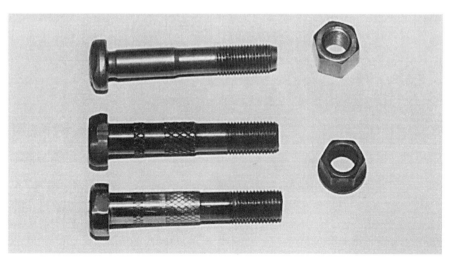

At top is factory-installed Nissan/Datsun 9mm rod bolt. At center is aftermarket 3/8-in. bolt for a big-block Chevy. Bottom bolt is aftermarket Chevy bolt with shank ground to exact size to ensure consistent press fit.

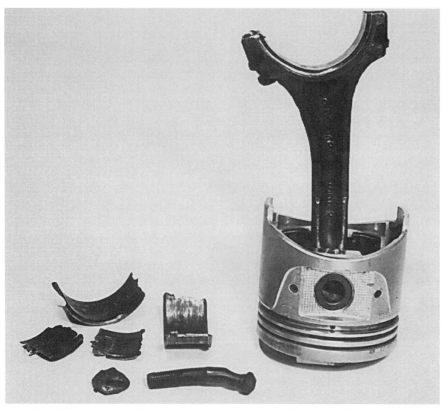

Results of failed rod bolt: destroyed connecting rod, "holed" block and scored crank (not shown). Use every precaution when choosing, inspecting, preparing and installing connecting-rod bolts. Tom Monroe photos.

Set of L16/L24 rods prepped according to instructions in text. Don Devendorf used similar rods in turbocharged 280ZX IMSA GTO car for two years with success before switching to factory sports-option rods to reduce preparation time.

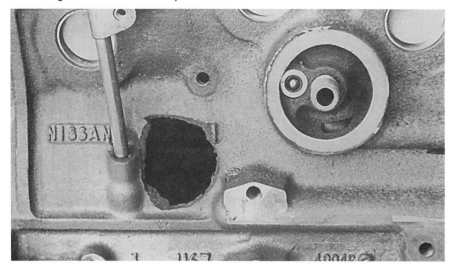

The diameter of the knurled shanks of the stock Chevy rod bolts is not consistent. Consequently, the press fits in the rods will vary. To correct this, polish the bolt shanks to a standard diameter. Then, ream the holes in the rods and caps. The rod holes should be reamed 0.0005-in. smaller than the shank diameter of the bolts so there will be a moderate press on the bolts. Ream the cap bolt holes to allow a slight interference or zero fit with the bolts.

Big-End Sizing—After you install either the new stock rod bolts or aftermarket bolts, size the rods, page 35. Remember to recheck the big-end bores after you get them back from the machinist to confirm they are right. They should be the desired size and perfectly round. Bore diameter will affect bearing *crush* and bearing oil clearance.

Small End—If you're keeping the standard pressed-in pins, doing any work at the small ends is probably unnecessary. However, check them to make sure. The small ends should be 0.0006—0.0013 in. *smaller* than the pins. But, if you are converting to floating pins, read on.

Center-to-Center Length—Check that big- and small-end-bore distance is within spec. This distance, or center-to-center length, should be within the specification listed in the chart, page 34. Or, if an aftermarket rod is used, it should agree with either the manufacturer's or your specifications.

Floating Piston Pins—Contrary to some opinions, it's OK to float the piston pins directly against the steel of the rods. But, if you prefer, you can bore the small ends oversize, then press in thin-wall

Connecting Rods

Nissan sports-option rod 12100-A7660 for L20B. Center-to-center length is 5.85 in. (148.6mm). Rods are no longer available.

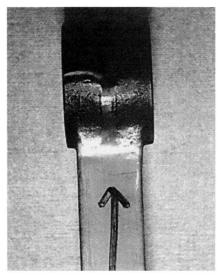

Even the "best" parts break. Immediately above arrow on Nissan factory sports-option rod is a crack across beam. Rods are no longer available.

NISSAN MOTORSPORTS CONNECTING RODS

Part No.	12100-A7660	12100-N7170	12100-A7602	12100-B2501	12100-R2000
C-to-C Length	148.6mm/5.85 in.	133mm/5.236 in.	140mm/5.512 in.	140mm/5.512 in.	140mm/5.512 in.
Bolt Part No.	12109-A7660	12109-A7660	12109-A7602	12109-B340	12109-R2002
Bolt/washer Part No.	12113-H5820	12113-H5820	12113-A7602	12113-A7602	12112-R2000
Piston-Pin Diameter	23.5mm/0.9252 in.	23.5mm/0.9252 in.	22mm/0.866 in.	23.5mm/0.9252 in.	22mm/0.866 in.
*Bolt Stretch/Torque	0.0051-0.0055 in.	0.0051-0.0055 in.	0.0047-0.0051 in.	0.0047-0.0056 in.	43-47 ft-lb

*In most instances 35-40 ft-lb of torque is required to obtain specified bolt stretch. If bolt does not "feel right" during tightening, replace bolt.

Connecting-rod 12100-R2000 costs about the same as a stock Nissan rod. It was originally used in the FJ20 four-cylinder, four-valve Nissan engine. This FJ20 engine was not used in any U.S. specification car.

bronze bushings.

Regardless of which way you go, the small-end bores must be honed to achieve the correct pin-to-bore *clearance,* not interference as with the pressed pins. Mike the piston pins, then hone the pin bores 0.0008—0.0011 in. (0.020—0.025mm) larger to give this clearance.

Caution: Don't forget to redrill the pin-oiling holes if you bush the small ends. Otherwise, the pins will seize in the rods in a heartbeat after you start the engine.

OPTIONAL FACTORY RODS

Although Nissan Motor Co., Ltd. (Japan) once offered several optional connecting rods through Nissan Motorsports, these rods are no longer available.

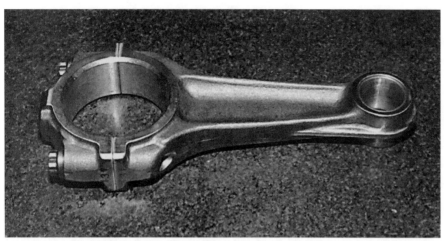

Rod from FJ24 Nissan/Datsun DOHC 2.44-liter race engine. This 5.51-in. (140mm) long rod can be used in some L-series engines. Big-end diameter is the same as L-series rods.

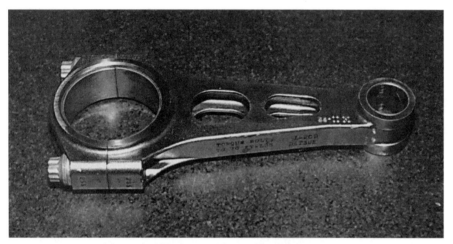

Investment-cast stainless-steel Mechart rod for L20B weighs only 560 grams. Rods are no longer available.

AFTERMARKET RODS

Several rod manufacturers make special rods for popular Nissan/Datsun racing engines. The most popular connecting rods come from Carrillo and Crower.

Carrillo rods are by far the most popular in the aftermarket. Their unique H-beam rods are forged, rather than machined from billet stock. Approximately 90% of Carrillo's rods are machined from the same forging.

Carrillo offers one rod with a 146mm (5.748 in.) center-to-center length, exactly to the L20B spec. Rods built for Nissan/Datsun engines are equipped with 3/8-in. SPS screw-in bolts. Carrillo recommends using aluminum/silicone-bronze bushings in the piston-pin holes. If an alternate-design rod is specified, an eleven-to-twelve-week delivery is required and at high cost.

Crower, a company known for their camshafts, crankshafts an other engine components, also produces custom billet rods. Crower designs the section width of the rod beam according to rod length and the application. This means they will build steel billet rods to blueprints or specified dimensions. Rod length, pin size, bolt size and beam dimensions are all tailor-made. Crower uses a unique beam design and big-end configuration that enhances the rigidity of the big-end bore.

One Last Thought—When considering which connecting rod to use in a race engine, remember that a broken rod can leave you with a basket full of junk. So, choose the right one. Then prepare it properly.

CHAPTER FIVE
Cylinder Head

John Olsen's 1997 200SX powered by an L16. One year earlier the same L16 powered Olsen's PL510 to a close second place at the SCCA National Championships. G. Hewitt photo.

The basic configuration of all standard Nissan/Datsun L-series heads is the same. They are all cast from aluminum and use a single spark plug for each combustion chamber and *non-crossflow* ports—both intake and exhaust ports are on the left side of the head. Wedge-shaped chambers and a single overhead cam (SOHC) are used. The cam is supported in four-cylinder and six-cylinder engines by four or five cam bearings or towers, for fours and sixes, respectively. Four- and six-cylinder heads are basically the same. As a result, L-series heads have many interchangeable parts.

An offshoot of the L-series engine is the LZ (Naps-Z) four-cylinder engine. As you may recall from previous chapters, this engine uses the basic L-series block, but is fitted with a unique cylinder head. This emission-control head is relatively efficient, with hemispherical combustion chambers and *crossflow* porting—intake and exhaust ports are on opposite sides of the head. Unfortunately, the small valves and ports are not conducive to most racing applications.

On the other end of the L-series engine performance scale are two *FIA* (Federation Internationale De L'Automobile) *Group 2* specification heads for four-cylinder engines.

Even though many of the cylinder heads for L-series engines are similar, specific features must be considered when choosing a cylinder head. A major consideration is combustion-chamber volume. Chamber volume (size) relates directly to compression ratio, which shouldn't be too high for street applications or too low for racing. As for other features, let's look at the various cylinder heads in detail.

STANDARD VALVES

Engine	Diameter (in./mm) Intake	Exhaust	Length (in./mm) Intake	Exhaust
L13	1.50/38	1.30/33	4.56/115.9	4.57/116.0
L16 early	1.50/38	1.30/33	4.56/115.9	4.57/116.0
L16 late	1.65/42	1.30/33	4.56/115.2	4.57/116.0
L18	1.65/42	1.30/33	4.53/115.2	4.57/116.0
L20A	1.50/38	1.30/33	4.36/110.7	4.36/110.7
L24 one 2-bbl	1.50/38	1.30/33	4.59/116.5	4.63/117.5
L24 up to 7-73	1.65/42	1.30/33	4.59/116.5	4.57/116.0
L24 after 8-76	1.65/42	1.38/35	4.59/116.5	4.57/116.0
L26	1.65/42	1.38/35	4.59/116.5	4.57/116.0
L20B	1.65/42	1.38/35	4,53/115.2	4.57/116.0
L28 before 7-80	1.73/44	1.38/35	4.53/115.2	4.57/116.0
L28 after 7-80	1.73/44	1.38/35	4.45/113.1	4.48/113.9

Unmodified stock combustion chamber of current SSS cylinder head.

FOUR-CYLINDER STREET PERFORMANCE

The SSS cylinder head is the most popular four-cylinder performance head. It has 40cc combustion chambers, 1.65-in. intake and 1.38-in. exhaust valves. The round intake ports are 1.5-in. diameter; exhaust ports are rectangular.

The original SSS head has been superseded several times by new designs. The current Nissan SSS head, 11041-U0600A, is best suited to 1600 and 1800cc street-performance engines. Using this head with its 40cc combustion chambers on a 2000cc-or-larger engine will probably result in excessive compression, depending on piston-to-deck clearance and piston-dish volume. Combustion-chamber volume is the same as in the stock L18 head and about 1.5cc larger than the stock L16 combustion chamber.

For some applications, such as an L18 with an aftermarket "hot-rod" camshaft, it may be desirable to install the L28's 1.73-in. intake valves, 13201-N4200. Breathing above 4000 rpm would be improved. Using a set of these larger valves requires installing new valve seats. However, this common machine-shop operation is relatively inexpensive.

Porting the SSS head is not required for street applications. There's enough "air" in the head as is. All that's needed is a small amount of polishing around the valve seats to break the machined edges.

If the compression ratio is too high, material can be removed from the combustion chamber. Do this by transferring the outline of the head-gasket *fire ring*—metal insert that circles the combustion chamber—to the head. Use the head gasket as a pattern to scribe its outline onto the head.

Remove material where it will be most effective—from around the valves to unshroud them. However, before attempting this modification, you'll need some special equipment and an understanding about what to do, see page 43. Use caution not to remove excessive material. The edge of the gasket must not overhang the combustion chamber. Also, the compression ratio will be reduced to an unacceptable level if too much material is ground away.

The currently available SSS head is modified from cylinder-head 11041-U8880. The intake ports of the standard 11041-U8880 head are bored from their original 1.375-in. diameter to 1.500 in. to make an SSS cylinder head. Consequently, the 11041-U8880 head is exactly the same as the SSS 11041-U0600A head, but with smaller intake ports.

The 11041-U8880 head is well suited to moderate-performance engines with relatively limited or stock carburetion. Bigger intake ports are great for making horsepower at higher rpm, but smaller ports give higher air/fuel-flow velocities for better low-end performance.

For 2000cc-or-larger-displacement L-series engines, use a standard L20B head, part number 11041-U6702. This head is identical to the SSS head except it has a larger chamber volume—45.2cc versus 40cc. This larger chamber is more compatible with the large *swept volume* of the larger-displacement engines.

An engine's *static*—calculated—compression ratio is the volume of one combustion chamber divided into the swept volume of one cylinder. Total combustion-chamber volume includes the volume at the cylinder-head chamber, head gasket, piston dish and area around the piston above the top ring. Most street engines have dished pistons, whereas most normally aspirated racing engines feature domed pistons. Because they project into the combustion chamber, domed pistons effectively reduce chamber volume, thereby increasing compression.

To determine compression ratio, use the following formula:

$$\text{Compression Ratio (C.R.)} = \frac{\text{Swept Volume (S.V.)}}{\text{Clearance Volume (C.V.)}} + 1$$

Where:

S.V. = engine displacement ÷ number of cylinders.

C.V. = cylinder-head volume + head-gasket volume + piston dish (or dome) + top-ring volume.

Modifications to the L20B head are the same as those I suggested for the SSS head: Z intake valves and cleanup around the valve seats. Or, because of the larger combustion chamber, it may be necessary to mill the head-gasket surface to raise the compression ratio.

The downside to milling the head and shimming the cam towers is that it puts the cam so high that getting a good wipe pattern on the rockers is difficult. Thicker lash pads will be required on all valves.

SIX-CYLINDER STREET PERFORMANCE

The early 240Z cylinder head, 11041-E3100, is the most popular head for use on L-series six-cylinder engines. It has 42.5cc combustion chambers, 1.65-in. intake and 1.30-in. exhaust valves, and 1.5-in. intake ports and rectangular exhaust ports. A useful modification to this head is to install 1.38-in. exhaust valves. It's also beneficial to install 1.73-in. intake valves with correspondingly larger valve seats.

The 240Z head should not be installed on some L28 engines. The result will be too much compression for the street. Some street enthusiasts have created real disasters by installing an *E3100* head milled "just a little bit" on a 1mm (0.040-in.) overbored L28 engine with flat-top pistons. This combination gives 11.3—11.5: 1 compression ratio. Try to find pump gas for this engine! Such an engine would be great for the budget racer, but not for the street, where the best you can buy is typically 87—91-octane pump gasoline.

Small combustion chamber of early 240Z head is a good choice for use on race engine where high-octane gasoline is available, but not for street engine. Valve-seat areas were machined for larger 280Z valves.

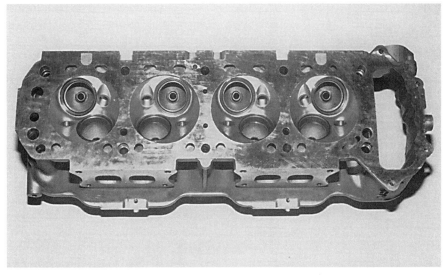

Crossflow Naps-Z head installed on 1980—83 four-cylinder Nissan/Datsun engines has small valves, cross-flow design.

For street performance, no porting is required and only a moderate amount of polishing. Polish the combustion chambers and around the valve seats to break the machined edges. That's all that is required. If combustion-chamber volume is too small, increase it slightly by grinding around the valves. Unshrouding the valves improves breathing and reduces compression to an acceptable level in one operation.

The 11041-E8803 cylinder head is excellent for use on an L28. This head is similar to the E3100 except it already has 1.38-in. exhaust valves and larger, 44.7cc combustion chambers. The E3100 head has the smaller, 42.5cc chambers.

PERFORMANCE MODS TO STOCK HEADS

Naps-Z—The Naps-Z head is used on all 1980-and-later passenger-car and 1981-and-later light-truck four-cylinder engines. For practical street application, the only way to increase compression with this head is to mill it. Higher-compression pistons are not generally available. And, installing larger valves is not practical because, in their stock

Twin-plug, hemi-chamber of Naps-Z head is desirable, but valve and port sizes restrict potential performance.

Naps-Z intake ports are small and positioned low relative to valve seats. Regardless of size, low port position results in small short-side radius that restricts port flow and, ultimately, engine performance.

Naps-Z exhaust ports are also small and positioned low relative to valve seats.

configuration, the valves nearly contact each other at TC because of valve overlap. Piston-to-valve clearance, however, is adequate.

Also, the Naps-Z ports are relatively small and there's not much room to enlarge them. But, unlike L-series heads, Naps-Z heads respond well to porting and polishing without removing much material. Typically, this consists of straightening the Naps-Z ports, and grinding away material at the bowl area and short-side radius above the valve seats to improve flow.

Standard L-Series—Performance modifications are easy to do on standard L-series cylinder heads. Valve sizes can be changed with ease. Also, the head-gasket surface can be milled to raise compression in moderate increments. Port and chamber work is easily done with a hand grinder on aluminum heads, particularly the L-series heads.

Matching the size, shape and location of the intake ports to the manifold runners and unshrouding the valves in the combustion chambers are crucial to good performance.

Match Ports & Unshroud Valves—To match the manifold runners to the cylinder ports at their mating faces, use the gasket trimmed to match the cylinder-head ports as a template. Once trimmed, the manifolds are installed with the gasket on the head and matchmarked. Use machinist's blue and a scribe to mark across the head, gasket and manifolds. Vertical and longitudinal positioning should be indicated. After marking, you can accurately position the gasket to the head for using it as a port-transfer template.

To unshroud the valves, start by transferring the outline of the head gasket you're going to use to the head. Position the gasket on the head and scribe around the chambers, using the fire-ring portion of the gasket as a pattern. Use caution not to grind past the gasket outline and undercut the head gasket. Polish the areas around the valve seats and the machined edges in the chambers.

Turbocharged or Supercharged Street Engines—The compression ratio for a *boosted*—turbocharged or supercharged—engine must be significantly reduced over what is acceptable for a normally aspirated engine. Typically, the compression ratio for a *boosted* street engine should be about 7.7—8.3:1.

As is the case for a normally aspirated engine, the first thing to consider when choosing a cylinder head for a turboed or blown engine is the static compression ratio. In some cases, there are heads that will give lower-than-original compression ratios. For instance, the L20B head, 11041-U6702, is probably best suited for boosted four-cylinder engines. The Naps-Z head is also acceptable. The turbo 280 ZX head, 11041-P9080, has the largest six-cylinder chamber volume at 53.6 ± 0.5cc.

Note: 1981-and-later 280 ZX valve stems are shorter than those in earlier six-cylinder heads by 2.5mm (0.100 in.). They were shortened to compensate for the combustion-chamber roofs being raised the same amount.

If, for some reason, you're using your engine's original head or an alternate and combustion-chamber volume is too small—compression is too high—all is not lost. The head can be modified.

As described earlier, chamber volume can be increased by hand grinding. In addition to this, other precautions should be taken when inspecting and preparing a head for a boosted engine. This is simply because of the higher combustion-chamber temperatures and pressures.

To ensure good head-to-block sealing, the head-gasket surface must be perfectly flat and smooth. Steel valve seats 0.070-in (1.78mm) wide should be used to increase valve-to-seat heat transfer. Combustion-chamber finish should be smooth and machined edges radiused. This will help reduce detonation by eliminating sharp, thin edges that would otherwise glow red and act like the glow plug of a model-airplane engine.

RACE-ENGINE HEADS

Selection—The choice of a cylinder head and modifications made to it for use on a race engine are critical factors to the engine's performance. Except for the FIA Group-2 four-cylinder head, covered separately, modifications to four- and six-cylinder heads for racing are basically the same.

The selection and modification process depends on several factors, such as sanctioning-body regulations, engine displacement and type of racing.

Sanctioning-body regulations are probably the most significant consideration during initial cylinder-head selection. Rules are simple, but absolute. For instance, if the cylinder head you use is illegal for one reason or another, your vehicle won't be allowed to compete. Or worse, if you do compete and your transgressions are discovered, you'll be dealt with severely. Also, you'll be suspect from then on, even though you correct the infraction. Therefore, restrictions on valve sizes, chamber modifications and piston-top configuration must affect head selection from the outset.

Within the rules, the addition of material to the cylinder head by welding may be an important option. The updating and backdating within the model series and the use of optional heads must be considered.

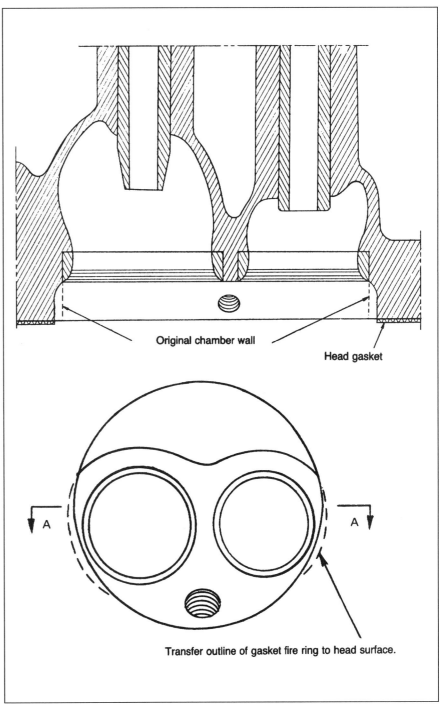

To avoid exposing underside of head gasket when unshrouding valves, transfer outline of head-gasket fire ring to head before grinding. Don't grind past dashed line indicating head-gasket fire ring.

The type of induction system allowed and used must be considered when determining cylinder-head intake-port size. The type of racing, engine-power

Pre-'78 L20B head 11041-U6702 with high-volume combustion chambers and rectangular exhaust ports is excellent choice for use on turbocharged four-cylinder engine. Tom Monroe photo.

Combustion chambers of late L28 head were welded to reduce volume by over 17cc.

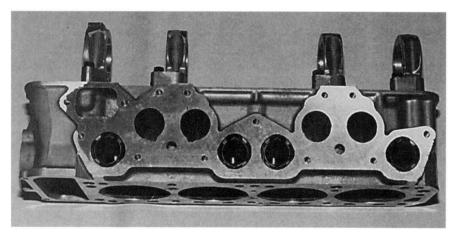

Post '78 L20B and post '77 L28 heads have round exhaust ports with steel liners. Compare exhaust ports to those of early L20B head pictured above.

band, maximum engine speed, and budget—size of your pocket book—also affect the selection of a cylinder head.

Engine displacement is a critical consideration on small-displacement engines because it can be extremely difficult to reach the desired compression ratio with a small swept volume. For this reason, FIA Group-2 and SSS heads are the two most popular four-cylinder race heads. Using an early small-chamber head on the L16, 1600cc engine may provide enough compression without welding.

Type of racing may narrow your choices. For instance, E3100 and E8800 cylinder heads are the most popular SCCA GT-2 category heads where heads cannot be welded. For IMSA or other applications where welding is allowed, virtually any head can be transformed into one with a desirable combustion-chamber configuration. Most six-cylinder heads have the rectangular exhaust ports that I prefer for racing. Some late 280Z heads have round exhaust ports with steel liners.

Steel liners in the exhaust ports of L-series heads are for reducing exhaust emissions, not increasing power. The liners are heated by the exhaust similar to the glow plug of a two-cycle model-airplane engine. The hot liners keep exhaust temperatures up to burn excess hydrocarbons as they exit the exhaust ports.

Exhaust-port liners are cast into the port walls, but they can be removed. Without the liner, each exhaust port is round and has a large volume. Also, the centerline and basic configuration of the round port is similar to the square port. Regardless, the square exhaust port remains the most desirable even though some racers use heads with round exhaust ports.

As discussed earlier, 1981-and-later 280ZX heads have the biggest chambers and a lower compression ratio because of their 0.100-in. higher roofs. Consequently, these heads are best suited for turboed or supercharged engines. A smaller chamber is preferred for a high-compression, naturally aspirated engine.

RACE MODIFICATIONS

Prior to starting any cylinder-head modification, options must be considered. Valve size, port size and shape, and chamber volume and shape must all be evaluated for the specific application. For instance, off-road racing or rally-car engines require small-diameter valves and ports to promote high-velocity air/fuel-mixture flow for enhanced low-rpm torque. On the other hand, road-race or drag-race engines make better use of larger ports and valve sizes to increase high-rpm power.

Welding ports to change their shape is possible. Although not a popular modification, it is feasible. For example, Norwin Palmer, a drag racer from Kansas, won the NHRA World Championship with a four-cylinder Nissan/Datsun L-series powered econorail. The huge 2300cc four-cylinder engine had an early FIA Group-2 cylinder head, 11041-22010. Palmer, a welding instructor by trade, raised the intake ports up close to the valve-cover-gasket surface. He raised the intake-port floors about 2 in. The resulting D-shaped intake ports protrude into the valve-train area so the welds are visible with the valve cover removed.

As previously discussed, compression ratio and modifications allowed to the combustion chamber must be considered. Such modifications range from grinding and surface milling, to welding.

Before I get into specifics concerning cylinder-head race modification, here's a suggestion. To achieve optimum port flow, cylinder-head modifications should be done by a professional. He'll have a flow bench and a lot of experience to back up his work. The professional will have seen the results—failures and successes—of different port configurations for various applications. To back this up, he'll have the documentation needed to reproduce the best ports for your specific application.

Cylinder-head porting services that specialize in Nissan/Datsun heads, such as B.C. Gerolamy in Sacramento, California, will give the best results. Gerolamy not only specializes in Nissan/Datsun heads, but also conducts ongoing development. He also markets cylinder-head porting kits through Cylinder Head Abrasives (listed in suppliers).

Now that I've suggested that you have a professional do your cylinder-head work, let's take a detailed look at cylinder-

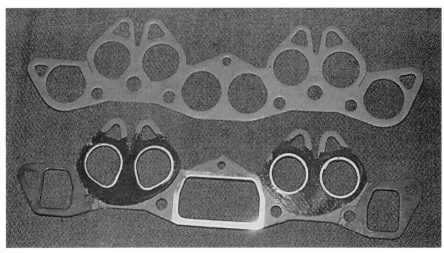

Gasket at top is for round exhaust-port four-cylinder head; other is for rectangular exhaust-port head. Note that exhaust ports are in same relative positions. Only their shapes changed.

Jim Fitzgerald used this E3100 head with intake ports modified to a D-shape.

head race modifications. You need to know the fine points of good head work whether you insist on doing the work yourself or will have it done, as I've suggested.

Port Modification—The main objective of modifying the ports is to increase airflow into or out of the combustion chamber. This is generally accomplished by enlarging and straightening the ports, producing a uniform port wall taper and maintaining a generous *short-side radius* —transition from the back side of the valve to the port floor. This work must be precise because the modified ports— intake or exhaust—must be identical when finished.

Attempting to grind a port for optimum

Floor of number-2 intake port in L20B head was ground through to water passage. Damage was repaired by welding.

Small hole (arrow) in number-3 intake port of SSS four-cylinder head was not discovered until after engine was assembled and run. Head should have been pressure-tested after porting work was completed. Water-passage hole—centered and below intake ports—was tapped for 1/4-in. pipe plug.

B.C. GEROLAMY COMPANY
2120 BLUMENFELD DRIVE
SACRAMENTO, CA 95815
916-922-7652

ENGINE L6 2.8 LITER / E 3100 HEAD
INTAKE 1.73-in. MANLEY
EXHAUST
INCHES OF WATER 25"
SCCA GT-2 Spec.
#6 SEAT

F/R	%	CFM
100	2/72.9	63.7	46.4
200	3/151	64.5	97.4
300	4/300	45.6	136.8
400	4/300	56.7	170.1
500	4/300	62.9	188.7
600	4/300	65.8	197.4
700	4/300	67.6	202.8

Flow chart for intake port of E 3100 L28 head modified for SCCA GT-2 competition: Valve lift, shown in first column, begins at 0.100 in., progressing in 0.100-in. increments, and ends at 0.700-in. lift. Column at far right is airflow in cubic feet per minute (CFM).

SSS four-cylinder head modified for racing by Slover's Porting Service. Intake-port diameter was increased to 1.625-in.

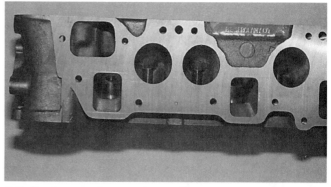

B.C. Gerolamy made extensive modifications to six-cylinder IMSA GTU head. Manifold bolt pattern indicates head is from late L28 EFI engine. Injector holes in the roof of intake ports were welded closed. Small hole below center of two intake ports at front of head is for dowel which aligns manifold and gasket to ports. Dowel is in the manifold.

ENGINE L6 2.8 Liter/E3100 Head
INTAKE
EXHAUST 1.380-in. Manley
INCHES OF WATER 25"

SCCA GT-2 Spec.
(standard rework)
#8 Seat

	F/R	%	CFM
100	2/76.6	50.0	38.3
200	3/160	45.5	72.8
300	3/160	61.0	97.6
400	3/160	71.2	113.9
500	3/160	85.2	136.3
600	3/160	90.8	145.3
700	3/160	92.4	147.8

Exhaust flow of E3100 L28 head: Flow follows 75% rule: It should be approximately 75% of intake flow.

ENGINE L6 2.8 Liter/N4258 Head
INTAKE 1.73-in. Electramotive Titanium
EXHAUST
INCHES OF WATER 25"

IMSA GTU Spec.
(welded chamber)
#6 Seat

	F/R	%	CFM
100	2/72.9	70.5	51.4
200	3/151	60.0	90.6
300	4/300	43.3	129.9
400	4/300	54.5	163.5
500	4/300	60.0	180.0
600	4/300	65.7	197.1
700	4/300	68.8	206.4

Intake flow of L28 N4258 head modified to IMSA GTU rules: Combustion chambers were welded.

ENGINE L6 2.8 Liter/N4258 Head
INTAKE
EXHAUST 1.450-in. Tilton
INCHES OF WATER 25"

IMSA GTU Spec.
(welded chamber)
#8 Seat

	F/R	%	CFM
100	2/76.6	53.1	40.7
200	3/160	50.8	81.3
300	3/160	71.1	114.7
400	3/160	82.6	132.2
500	3/160	84.8	135.7
600	3/160	86.5	138.4
700			

Exhaust flow of L28 N4258 head modified to IMSA GTU rules: Head is same as for intake-flow chart, above right.

ENGINE L6 2.8 Liter/P7900 Head
INTAKE
EXHAUST 1.50-in. Electramotive Titanium
INCHES OF WATER 25"

IMSA GTU Spec.
(welded chamber)
Round Exhaust Port

	F/R	%	CFM
100	2/76.6	55.1	42.2
200	3/160	50.8	81.3
300	3/160	65.0	104.0
400	3/160	74.4	109.0
500	3/160	78.0	124.8
600	3/160	80.8	129.3
700	3/160	85.0	136.0

Exhaust flow of L28 P7900 head with round exhaust ports modified to IMSA GTU rules: Compare flow figures to those of N4258 head at left.

Gerolamy-modified head exemplifies true attention to details. Note how port, seat and chamber blend smoothly.

Close-up view of valve seat and combustion chamber: B.C. Gerolamy cut valve seat with Hall-Toledo® orbital grinder to give smooth-radius valve seat. Note matte finish from lapping valve.

flow on a head that you don't have any experience with can be risky. Even if you don't *strike water*—grind into the water jacket—you may grind a port wall so thin that it will eventually crack or leak coolant. Therefore, if you've never done porting on a Nissan/Datsun L-series cylinder head, be conservative. Don't start out by removing large amounts of material.

Instead, concentrate on straightening each existing port and blend it into the *bowl*—area immediately behind the valve seat. The port should taper down from the *mouth*—entrance at manifold-mounting surface-into the middle, then enlarge slightly as it enters the bowl. This taper—transition from large to small, and to large again—must be gradual and smooth. The radius on the short side or floor of the port just above the valve seat is critical to airflow. If the short-side radius is too large, it will *kill* airflow. To make things more difficult, the short-side radius is difficult to form.

The intake ports can be safely enlarged. Start by raising the roof about 1/8 in. Then, remove about 1/16 in. from the side walls while concentrating on straightening the port. To produce a round port that is consistent in size, utilize a selection of miscellaneous valves as templates. These valves, with different heads, are used for checking port size at different locations by inserting valve into each port a given depth. This ensures that the port cross section is a certain size at a given distance from the port mouth. Cross sections of each port can be checked along its full length by utilizing a combination of these valve templates.

Port finish is important for achieving maximum airflow through a given port once the optimum shape is obtained. The surface should be as smooth as possible. In fact, a mirror finish is best. A number of different grits and shapes of sanding rolls will be required to achieve such a finish.

Generally, the ports should be finished with 100-grit sandpaper rolls. The corners of the exhaust ports and some areas in the intake ports will require tapered or cone-shaped sandpaper rolls. Companies such as Gerolamy's Cylinder Head Abrasives can supply you with the needed porting "kit."

Valve Seats—Valve seats and valve heads are probably the most neglected area when it comes to proper detailing. This is an area where minute details count.

Unlike many production engines that have only a 45° angle at the valve seat, Nissan/Datsun engines use a multi-angle seat. Such a seat, often referred to as a *three-angle valve* job, is standard on all L-series intake-valve seats and many exhaust seats.

In addition to the 45° seat, there are the top and *bottom cuts*. The top cut on the combustion-chamber side of the 45° seat, is 30° to horizontal. The bottom cut—on the port side of the seat—is 60° to horizontal. These angles are in addition to the 90° vertical—*throat*—and the 0° top plane of the valve-seat insert.

Compared to a single-angle seat, airflow is better over a multi-angle valve seat. Airflow can be further improved with a *radiused* seat. In effect, a radiused seat is made up of an infinite number of angles, starting from the 90° throat angle and blending smoothly to the 0° face of the seat insert. Rather than the valve face seating on a wide band, normally a 45° seat, its seat-contact area is a narrow line. However, the radiused seat is not perfect. Because its valve-to-seat contact area is narrow, durability is not as good as with the traditional 45° seat.

The finish work required to do a valve seat correctly requires the professional touch. For instance, a cylinder head from B.C. Gerolamy is a true work of art. The attention to detail shows up in how the port blends smoothly into the seat, the finish of the seat, and the way the seat blends into the combustion clamber.

Valve Job—The *valve job* on any race engine is crucial to its performance. Let's look at the seats first, then the valves.

Seat shaping: The 45° intake-valve seats should be 0.030—0.040-in. wide and exhausts 0.040—0.050-in. wide. The top edge of the seat should contact the valve face about 0.015—0.020 in. behind or below the face OD or margin.

To lower or reduce the diameter of the 45° seat, cut a 30° angle above the seat. Raise or widen the seat by doing additional cutting on the 45° seat angle. Once all face/seat work is completed, narrow the seat by cutting a 60° or 70° angle at the bottom, or inside, edge or the seat. Seat angles should then be carefully blended. Also, the valves should be lapped to ensure a perfect seal. Lapping sticks and fine-grade lapping compound are available at most auto-parts stores.

Valve depth: Prior to narrowing the seats, check valve *depth*—the location-of the valve in the combustion chamber relative to the gasket surface. Valve depth increases with valve seat-to-gasket-surface distance. Consequently, valve depth is seat depth. The depth of all valves—seats, if you prefer—should be the same.

Valve depth can be measured using a depth micrometer. Distance is measured from the head-gasket surface to the valve head. To ensure accuracy, a steel ball can be epoxied to the exact center of one intake-valve head and another ball to an exhaust-valve head.

To measure valve depth, the intake or exhaust valve is dropped onto a seat. Then, its depth is measured using the depth micrometer. Record valve-depth measurements as you go for each intake and exhaust valve. Once you've checked both intake- and exhaust-valve seats, cut down the seats to match the deepest one.

With all valves set at the same depth, lash-pad selection, piston-to-valve clearance and valve-spring installed-height adjustments are simplified. Equal-depth valves also minimize volume changes between combustion chambers.

Valve-head shapes: Not only does seat contour affect flow in and out of a combustion chamber, so does valve contour. In

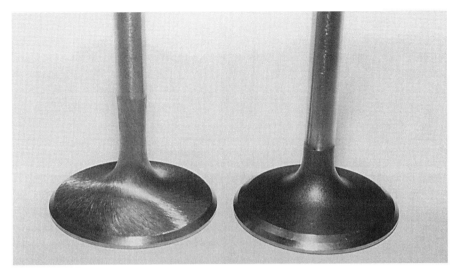

Valve at right is stock 1.73 in. L28 intake valve. Valve at left is same valve modified, available from Nissan Motorsports, part number 99996-N1100.

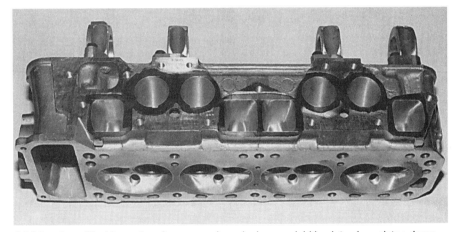

SSS head modified for racing: Area around spark plug was laid back to clear piston dome.

fact, valve shape has a significant affect on flow in the bowl area and over the seat.

Starting with the basics, the 45° valve face should be 0.060—0.070-in. wide. The valve-head *margin*—width between the valve-face OD and valve head—should be 0.030—0.040 in. The valve face is narrowed and blended into the underside of the valve head by *back-cutting* it at a 30° angle. The angles should be blended using other angles to smooth the transition between the 45° face and 30° back-cut.

Nissan Motorsports markets modified stock valves that already have the heads modified. They are machined in a numerical-control (NC) lathe to ensure that all intake and exhaust valves have the identical profile. The stem is narrowed just under the head and up to where it operates in the guide. This lessens the disruption in airflow caused by the stem diameter. The underside of the valve head is polished smooth and the edges of the valve face and margin are radiused and

To ensure modified engine will have desired compression, combustion-chamber volume must be measured. This requires a plate for sealing chamber flush with gasket surface and a burette for measuring liquid volume.

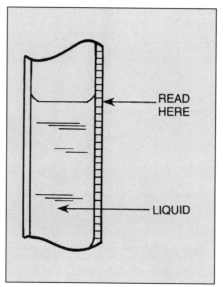

When measuring liquid volume with a burette, read graduation that aligns with bottom of meniscus-surface of liquid.

polished. The top side of the valve heads are *tuliped*—hourglass shaped —to the same profile and are also fully polished.

Aftermarket valves, such as those from Manley, also have valve heads shaped to enhance flow. There are all types of configurations and opinions of what the ideal valve-head shape should be. Another factor to consider is flow direction. For instance, the ideal shape for an intake- and exhaust-valve head is significantly different.

When you work with an experienced head porter, that person can either supply or recommend valves to produce the best flow for your application.

COMBUSTION CHAMBER

The shape of the combustion chamber and cylinder wall in the immediate area of the valve affect airflow. Also, the piston dome, especially any part of it that protrudes above the valve margin during overlap, generally upsets airflow.

To modify a combustion chamber, start by unshrouding the valves. Transfer the outline of the head-gasket fire ring to the head-gasket surface. Don't go beyond this line when removing material surrounding the valves. Radius the area around the spark-plug side of the chamber into the gasket surface.

Some engine builders grind the spark-plug area of the chamber back to the gasket outline. They then use a piston with a dome that protrudes into this relieved area. However, many engine builders and cylinder-head experts, such as B.C. Gerolamy, don't grind the spark-plug area of the chamber. Instead, they simply polish this area and use a modified piston dome that fits into the chamber.

To unshroud the area around the valves on the side away from the spark plug, merely radius the edge where the chamber meets the gasket surface. Any unnecessary material removal won't necessarily improve breathing, but will reduce compression.

Attention to detail is especially crucial in the area near the valve seats. The surface should be smooth and blended to aid airflow. Edges created by valve-seat cutters or stones, machining marks, and incorrect radiuses will have detrimental effects on airflow. If new seats are installed, they must be driven flush with the chamber roof. Also, the top radius of the seat, the point where the top of the seat angle blends into the chamber, must be smooth.

The end of the spark-plug threads should be, when installed, flush with the chamber surface. This can be accomplished by interchanging copper spacers until you find the one that sets the plug to the desired depth. Copper plug spacers are available in different thicknesses. Nissan Motorsports offers 0.096- and 0.135-in.-thick washer-type spacers.

The threads in the spark-plug hole should be detailed. For instance, spark-plug threads may be exposed at the combustion-chamber end of the hole. These threads must be polished smooth to prevent them from causing preignition.

Adjusting Chamber Volume—The final combustion-chamber modification is the measurement and adjustment of chamber volume. On most naturally aspirated race engines, the smallest combustion chamber is desirable to achieve maximum compression ratio. The other important consideration is to ensure that all chambers have the same volume. To accomplish this, you must measure and record the volume of each chamber.

Volume-Measuring Tools—To measure combustion-chamber volume, you'll need a 4-in.-square, 1-in.-thick piece of Plexiglass and a chemist's *burette*—graduated glass tube with a stopcock at the bottom. The Plexiglass plate should have a 1/4-in.-diameter hole that's chamfered at both ends. The burette should have a 100cc capacity with 0.1cc graduations and the zero, or full point, at the top.

Install the intake and exhaust valves with a coating of grease on the faces to provide a watertight seal. The spark plugs should be installed with the correct-thickness copper spacers. Support the cylinder head so the gasket surface is horizontal. Surround the combustion chamber on the gasket surface with a bead of grease and position the Plexiglass plate over the combustion chamber. Push down on the

Cylinder Head

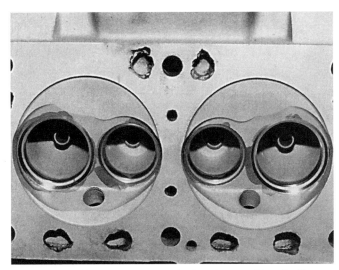

Erosion around water passages of otherwise good head will soon cause serious problems if not corrected. Such damage can be repaired by welding without causing serious warpage. Regardless, head must be resurfaced afterward to smooth welds and ensure straightness.

Engine over-rev and subsequent exhaust-valve contact with piston led to exhaust-valve head breaking and severe damage to head. As you'll see, such damage is repairable. See damage to piston, page 16.

Damaged head was stripped and cleaned, and damaged area ground and polished clean prior to welding. While we were at it imperfections in combustion-chamber-1 were corrected.

Head was welded by a casting mold-repair company. As you can see, the welder was very good at welding cast aluminum.

plate so it's flat against the gasket surface. Check the grease seal at the periphery of the combustion chamber. If it's fully *wetted* to the plate, the seal is good.

With the stopcock closed, fill the burette past the 0 mark with water or clean solvent. The liquid should be bubble-free. Hold the burette vertical and zero the liquid. Drain fluid from the stopcock until the bottom of the *meniscus*—U-shaped surface of the liquid—lines up with the 0 mark.

To measure the combustion-chamber volume, drain liquid from the burette into the combustion chamber through the hole in the Plexiglass plate. Tilt or rock the head to vent any bubbles from the fill hole. Once you've completely filled the combustion chamber, check how much fluid was drained from the burette. Read chamber volume directly from the burette. Again, read from the bottom of the meniscus. Record this volume, then repeat the process on the each of the remaining chambers.

All chambers must be enlarged to match the volume of the largest one. Maximum allowable chamber variation is 0.5cc. Chambers can be enlarged by doing additional grinding, but all should be the same shape. As you should know, chamber volume is used to determine compression ratio.

CYLINDER-HEAD REPAIR

Aluminum heads are relatively easy to repair. And, compared to most other overhead-cam cylinder heads, repairing warped Nissan/Datsun L-series heads is easier.

Top and bottom surfaces were machined to ensure a perfectly straight head. Head was warped only 0.004 in. Note fine finish left on the gasket surface to ensure a good gasket seal.

Combustion chamber is partially machined after new seats are installed. Valve seats are cut in this operation. Remainder of combustion-chamber repair will be done by hand grinding.

This doesn't mean OHC heads with integral cam bearings can't be milled. Although the head-gasket surface of heads with integral cam bearings can be straightened by milling, the bearing-bore center line will remain out of alignment. The result is that the head must be replaced.

As for an L-series head, the cam towers can be removed and the top and bottom surfaces of the head milled. This gives a flat surface for remounting the cam bearings. Bearing-bore alignment is restored. Cam-to-crank distance is restored by inserting shims between the head and cam towers.

This feature is significant because aluminum-head repair work typically involves welding. And, any significant amount of welding on a head will cause warpage. This is not a serious problem with L-series heads.

Repair Welding—Prior to welding a head, strip it completely. This, of course, includes removing the cam towers, dowels and pivot bushings. Once stripped, thoroughly clean the head. Grind and polish the area(s) that need to be welded. This is a must.

Clean Head—All traces of impurities related to combustion, such as carbon, must be completely removed prior to welding. If they are not, the remaining impurities will cause porosity in the weld, which will be obvious after the welded area is milled or ground. If not discovered and corrected, porosity can lead to combustion, coolant or oil leaks and further cracking.

Bend Head—If a significant amount of welding is to be done or if only a little amount of material can be milled to straighten the head after welding, it is best to bend the head *backward*—up at the ends. After bending, the head-gasket surface will be *convex*. This prebend will compensate for warpage that bends the head in the opposite direction—*concave*—that occurs when the combustion chambers are welded. The amount of preload or bend depends on the location and amount of heating or welding.

Bolt the head upside down to the deck of the stripped block. Support the head off the block with a spacer block at its center. The spacer should span the width of the block, be approximately 1 in. wide, and at least 0.100 in. thick. Secure each end of the head to the block with two short head bolts. Gradually tighten the bolts and check the bow in the head with a precision straightedge and feeler gages.

You'll need a 36-in.-long straightedge for checking a six-cylinder head; 24 in. will do for a four-cylinder head. With the straightedge flat against one end of the head, check the gap at the other end with your feeler gages.

How much prebend to put in a head is subjective. It depends on how much welding will be done. This is something you and your welder must judge. Use the following as a guide: For small to significant amounts of welding, prebend 0.020—0.040 in. for a four-cylinder head and 0.020—0.060 in. for a six-cylinder head, respectively; 0.020 in. if there's a minimal amount of welding to be done and 0.040—0.060 in. if you're going to do a lot of welding.

If you guessed right, your weld-repaired

Cylinder Head 55

Combustion-chamber repair is finished. Combustion chambers were welded to increase compression and improve chamber shape when head was originally modified. Head is not legal for SCCA Club Racing, but I don't understand why.

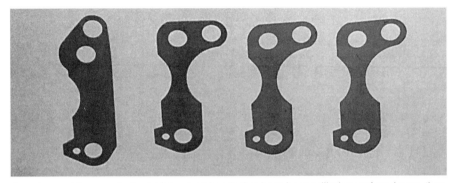

Cam-tower shims are designed for use on heads that have been milled or surfaced more than 0.030 in. Stacking as many as three shims per tower is common on heads milled 0.100-in. thinner than 4.218-in.-thick stock head. Shims are typically 0.030 in. thick.

Cam towers are installed and checked for alignment with cam. If cam turns freely, bearing bores are aligned. If not, mounting bolts are loosened and retightened after striking tops of towers with mallet and rechecking cam rotation. This is done until cam rotates freely and all mounting bolts are torqued to 10—13 ft-lb (1.4—1.8 kg-m). Tom Monroe photo.

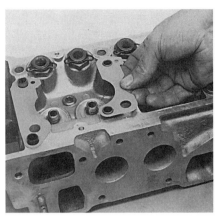

After locating dowels are in place, shims are installed as required. Tom Monroe photo.

head will only show 0.005—0.008-in. warpage after it has cooled. Use your straightedge and feeler gages to check. Whatever is required to clean up the head, the same amount must be milled from both sides.

Shim Cam Towers—After the head is milled, shims are installed between the cam towers and head if more than 0.030-in. total is removed from the head. This restores cam-to-crankshaft distance. Standard cylinder-head thickness—distance between top and bottom surfaces—is 4.218 in. Special head-saver shims for shimming the cam towers are available from most auto-supply stores.

Welding Combustion Chambers to Increase Compression—As mentioned previously, the head must be absolutely clean before you can weld the combustion chambers. This applies to both old *and* new heads. Polish the chambers with cartridge rolls, then with a rotary wire brush.

Bend the head backward the maximum amount because welding all combustion chambers is a significant amount of welding. Therefore, using my recommendations, a four-cylinder head should be bent backward about 0.040 in. and a six-cylinder head about 0.060 in.

Instruct the welder to "jump around," or weld a small amount in one area at a time. This will keep welding heat from being concentrated in one place, thus helping to reduce warpage and prevent cracking.

Ideally, the head won't exhibit more than 0.010-in. warpage after a significant amount of welding is done. Regardless of how much warpage occurs, you'll have to mill *both* sides of the head to make it straight and the top and bottom surfaces parallel. Also, all valve seats must be replaced. Welding and the resultant distortion will affect the press-fit of the seats. Also, the valve-seat bores may no longer be round or correctly sized. So, they must be remachined.

Straightening Head—Extreme engine overheating generally causes the head gasket to blow and severely warps the cylinder head. If a race engine is run for long at high speed and load with a head gasket that's blown between combustion chambers, the high-temperature combustion leak may burn a *trough—notch*—in the head at the gasket surface between adjacent chambers. Before straightening, the trough will have to be welded.

However, if the head is "only" severely warped, it can be corrected by milling the top and bottom surfaces. As a rule, if the head is warped no more than about 0.015 in., it's OK to mill both sides without straightening. Note that warpage requiring the removal of as much as 0.030 in. per side is unacceptable for most race engines.

To repeat, if the top and/or bottom surfaces of the head were machined more than 0.030 in. (0.76mm), restore the cam-sprocket-to-crankshaft distance by installing shims between the cam towers and cylinder head.

If you're machining a warped cylinder head, first measure its thickness at the front, middle and rear. Remember, standard head thickness is 4.218 in. After receiving the machined head, remeasure head thickness at the same three points. Any difference will be the amount of material removed by the machinist. Removing too much material may cause piston-to-valve interference.

Bend Head—If the head is warped

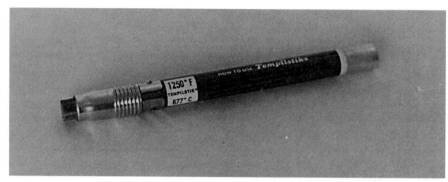

Temperature indicator will melt when surface it's applied to reaches indicated temperature. Temperature on Tempilstik indicator here is too high for aluminum at 1250F (677C). Instead, get indicators rated at 300F (150C), 325F (165C), 350F (175C) and 375F (190C). Tom Monroe photo.

more than 0.015 in., it should be straightened. This is done by bending the head backward, or opposite the warpage, and heating it evenly. The head is then allowed to cool while still preloaded.

Begin the bending process by totally stripping and thoroughly cleaning the head. Now, place the head upside down on top of the stripped block. Support the head in the center with a spacer block (see page 54), then secure the head with two short head bolts at each end. To bend the head backward, tighten the bolts until the head is bent backward about 1.5 times the amount that it is warped. Again, use feeler gages and a precision straightedge to measure the amount of reverse bend applied (see page 54).

Evenly heat the entire head to 350F (177C). Do this with an acetylene torch equipped with a *rosebud* tip. The rosebud tip is used because it puts out a lot of heat, and the aluminum head requires a lot of heat. So, make sure you have plenty of acetylene and oxygen in your bottles before you start.

To ensure that the head is evenly heated, you'll need several temperature-indicating crayons or bottles of temperature-indicating paint. The crayon or paint is applied to the surface of the head in several locations, including the ports, combustion chambers and valve-train side. This is done to ensure that the head is evenly heated. When the indicated temperature of the crayon or paint is reached, it melts like wax on a hot surface and will smear when rubbed.

Temperature-indicating crayons or paints are available from welding-supply stores. Tempil® and Omega Engineering® are two brands that come to mind. You'll need 300F, 325F, 350F and 375F temperature indicators to check head temperature as you heat it.

Once the entire head is heated to 350F, which may take up to 30 minutes of heating with your rosebud-equipped torch, allow the head to cool while it remains bolted to the block. To be certain that you've given it enough cooling time, allow the head to sit overnight.

After it has fully cooled, remove the head from the block and check for straightness. Ideally, it should be extremely close to straight, or within 0.005—0.007 in. If this is the case, both sides of the head can be milled slightly to ensure the top and bottom are parallel and perfectly straight. If the first try doesn't get the head within 0.005—0.007-in. limit, it's OK to repeat the bending process as needed.

Cylinder Head 57

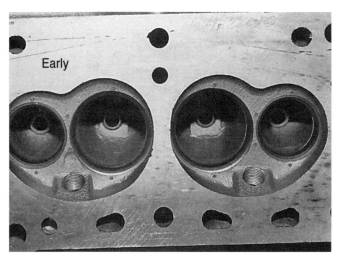

Intake valves of early FIA Group-2 head 11041-22010 are too large relative to exhaust valves.

Later FIA Group-2 head has smaller intake valve, larger exhaust valve and 2cc-larger combustion chamber.

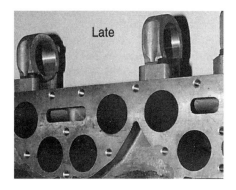

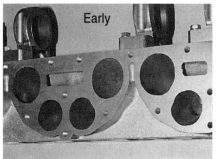

Head at bottom is early-version FIA Group-2 head 11041-22010. Head at top is later version of head, part number 11041-N7120. Compare port sizes. Smaller intake ports of later head are more desirable.

A race-engine cylinder head must have the top and bottom surfaces exactly parallel and perfectly straight. The surfaces must also be smooth, especially the head-gasket surface. It should have the smoothest possible finish. In fact, consider finishing the head-gasket surface on a *lapping plate*. Here, the gasket surface is polished on a precision surface plate with lapping compound. The object is not to remove material, but to give the head-gasket surface a near superfinish.

Replacing Valve Guides & Seats—

Guides and seats with oversize ODs and steel intake-valve seats are readily available. Brass intake-valve seats that were originally installed in some L-series heads should be replaced with steel seats. All Nissan/Datsun L-series heads come stock with cast-iron guides, which are typically replaced with bronze guides.

Most brass intake-valve seats are not acceptable for racing. These seats are too soft to endure the pounding that results from the combination of high-load valve springs and radical cam profiles. In fact, in totally stock engines it's not uncommon to find brass seats that are severely sunk after 75,000 easy street miles.

SSS four-cylinder heads and all L28 heads are originally equipped with steel intake-valve seats. The larger 1.73-in. intake- and 1.38-in. exhaust-valve seats can be easily installed into heads with smaller valves and seats, see pages 42-43.

Steel valve seats may be inadequate on high-output turbocharged engines. Exhaust temperatures reaching 1800F are excessive for acceptable valve-seat life, but such temperatures may exist in a turbocharged race engine. If this is the case, there is an alternative, valve seats made of beryllium copper. These seats will endure the extreme heat encountered in a racing-turbo application.

Any good engine machine shop can replace press-in valve guides and seats.

FOUR-CYLINDER FIA GROUP-2 HEADS

The most popular racing heads for Nissan/Datsun L-series four-cylinder engines are the FIA Group-2 versions. The FIA Group-2 head 11041-22010 has been superseded by the *late* version 11041-N7120. The *early* Group-2 head was first available in the USA in about 1970. Around 1980, it was superseded by the late version, but at five to seven times the cost of the early head.

A full-race-prepared SCCA GT-3 spec L20B engine will produce 8—10% more power with an early FIA Group-2 head as compared to the same L20B equipped with an SSS head. Additionally, the late-version FIA Group-2 cylinder head will result in slightly more power than an engine equipped with the early head. Most

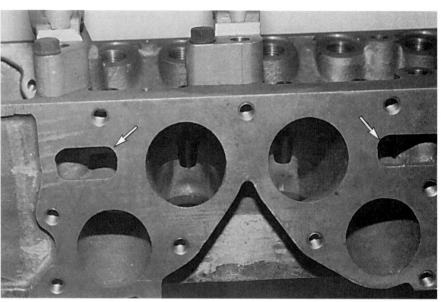

Early FIA Group-2 head has cast-iron cam towers with bushing inserts. Water passage over the exhaust port (arrow) routes coolant from head into a collector log at top of intake manifold.

Round intake and exhaust ports are approximately the same size in later-version FIA Group-2 head. Straightness of high intake port significantly aids intake airflow. Arrows indicate water passages. Intake valves are smaller than early FIA head; exhaust valves are larger. Standard aluminum cam towers are used.

FIA GROUP-2 CYLINDER HEADS

"EARLY" CYLINDER HEAD 11041-22010
- Intake-valve diameter: 45mm (1.77 in.)
- Exhaust-valve diameter: 35mm (1.38 in.)
- Valve-stem diameter: 9mm (0.354 in.)
- Combustion-chamber volume: 38.0cc
- Cast-iron cam towers with bushing inserts
- Special rocker-arm pivots
- Special valve springs, retainer and keepers

"LATE" CYLINDER HEAD 11041-N7120
- Intake-valve diameter: 44mm (1.73 in.)
- Exhaust-valve diameter: 36mm (1.42 in.)
- Valve-stem diameter: 8mm (0.312 in.)
- Combustion-chamber volume: 40.0cc
- Standard cam towers
- Standard rocker-arm pivots and hardware
- Standard valve-spring retainer and keepers

Part Description	PART NUMBER "Early"	"Late"
Cylinder head	11041-22010	11041-N7120
Intake valve	13201-22010	13201-A7660
Exhaust valve	13202-22010	13202-A7660 (E4621)
Outer valve-spring seat	13205-22010	13205-N3120 (E4621)
Inner valve-spring seat	13206-22010	13206-N7120
Valve-stem seal	13207-22010	13207-A7660
Keeper	13210-22010	13210-N7120 or 13210-0100

of this power increase is due to valve-size and port differences. In the late Group-2 head, the intake valve is 1mm smaller and the exhaust valve is 1mm larger. The good news is that the smaller intake valves and larger exhaust valves can be installed in an early head.

In addition to the valve sizes, other changes were made to the late head. To best describe the early and late FIA Group-2 heads, following are features of each:

Both versions:
- Fit all four-cylinder L-series cylinder blocks.
- Use standard head gaskets.
- Use standard head bolts.
- Use standard camshaft and rocker arms.
- Use standard cam sprocket and cam-retaining plate.
- Use standard-type spark plugs.
- Use same special exhaust header.
- Use similar, but different, special intake manifolds.

The early FIA Group-2 head, 11041 22010 was a significant departure from the stock cylinder-head design. Also, when compared to the late-version 11041-N7120 head, the early head is considered "overkill." For example, the early head has cast-iron cam towers with insert bushings, larger valve stems and special rocker-pivot assemblies. As you can see from the accompanying chart, the basic design of the late head is similar to the design of the stock L-series head. The late head utilizes all stock-size components except valve-head diameters and lengths.

Besides hardware differences, other significant differences between the two FIA Group-2 heads are port and valve sizes. According to B.C. Gerolamy, the intake port and valve size on the early head are too big. They reduce intake-air velocity, which adversely affects the engine's ability to accelerate, especially at low speeds. On the exhaust side, the early exhaust port and valve were too small. On the other hand, he considers the late head to have the ideal intake and exhaust-port and valve-size combinations.

Gerolamy goes on to say that he has significantly improved the performance of the early-style FIA Group-2 head. In some instances he has used the 1.73-in.-intake and 1.42-in.-exhaust valves in the early head. In addition to the small intake and larger exhaust valves, Gerolamy gained flow through the exhaust port by increasing its size.

OTHER OPTIONAL L-SERIES CYLINDER HEADS

Nissan Motor Co. Ltd. (Japan) offered two four-valve heads for four-cylinder L-series engines. An independent manufacturing firm in Japan offers a four-valve head for L-series sixes, but only a two-valve, crossflow head is available for the sixes from Nissan Motor Co.

Optional Four-Cylinder Heads— The two four-valve race heads produced by Nissan (Japan) are basically the same. Only combustion-chamber volumes differ. Small-chamber heads were meant for 1600cc Formula Pacific race engines. The larger-chamber heads were for 2000cc FIA Group-4 World Rally Championship engines and FIA Group-4 and -5 road-racing engines.

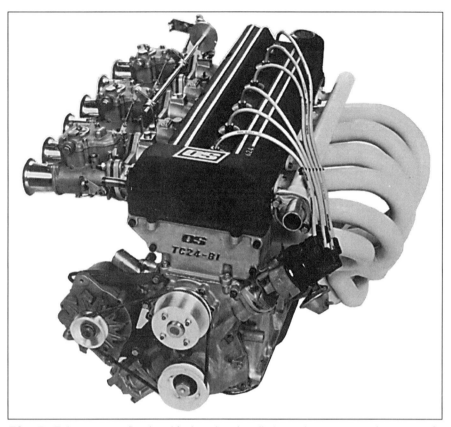

TC24-B1 Twin-cam crossflow head for L-series six-cylinder engines was manufactured by O. S. Giken in Japan. You'll have to go to Japan for this one. See also exploded view on page 60. Photo courtesy O. S. Giken.

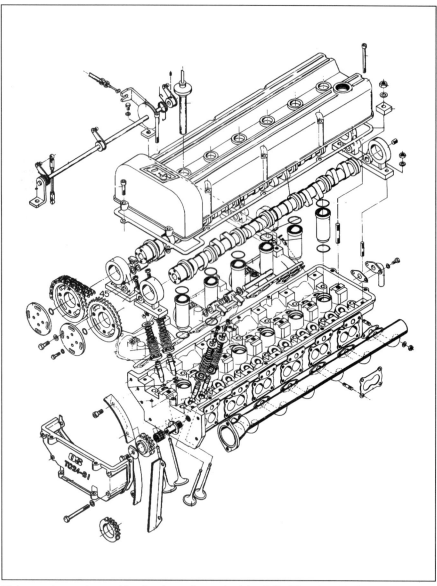

Unique valve and cam-drive hardware is required for O.S. Giken head TC24-B1. Drawing courtesy O.S. Giken. See photo, page 59.

A few four-valve race heads have found their way into the USA on complete engine assemblies. It's possible to acquire them through one of the many Japanese engine suppliers—junkyards or used-parts vendors.

An engine equipped with one of these heads will produce 15—20% more horsepower than a comparable engine with a two-valve, non-crossflow cylinder head. However, both the horsepower increase and high cost of using these heads have resulted in most race organizations disallowing their use.

Optional Six-Cylinder Heads— The optional two-valve, crossflow six-cylinder head is *not* available through Nissan Motorsports USA. The reason for this is it's not allowed in many racing classes. However, this isn't a major loss because the crossflow head will give "only" a 4—6% horsepower increase over a comparable two-valve, non-crossflow head. Consequently, cost versus potential power increase doesn't make this a highly desirable cylinder head.

Some race engines with this head found their way into the U.S. Bob Sharp raced a 280Z IMSA GTU car with this cylinder head in the late '70s.

A six-cylinder four-valve head, called *TC24-B1,* is manufactured by O.S. Giken, a private firm in Tokyo, Japan. One of these cylinder heads was brought to the U.S. on a rally-spec 280 ZX by a Japanese team. The car was campaigned at the '82 SCCA Reno Rally. The car with cylinder head intact was returned to Japan.

The TC24-B1 head is not available in the U.S. It is, however, available directly from O.S. Giken in Japan for about $12,000. If you're interested, the address is 464 Okimoto, Okayama, Japan.

New four-valve race heads have never been available directly from Nissan Motorsports USA. However, if you can get them, both versions are enormously expensive at around $10,000. To further escalate the cost of using these four-valve heads, a special cam-drive setup is required. Also, special pistons with domes and valve pockets that conform to the four-valve combustion chamber must be used.

CHAPTER SIX
Camshaft & Valve Train

Spencer Low won 1984 and 1985 HDRA Class 7S Championship in King Cab Nissan.

All Nissan/Datsun production L-series and Naps-Z engines have a single overhead-cam (SOHC) valve train. This chapter covers the valve train used with the non-crossflow L-series head, not the later Naps-Z head.

In the L-series valve train, four-cylinder heads use four cam towers to support the cam. Five towers are used in sixes. Rather than the cam lobes running directly over the valve-stem tips, rocker arms operate between the cam and valves. One end of each rocker arm rests on a pivot, the other on the tip of a valve. The cam lobe rotates against the top side of the rocker arm between the pivot and valve tip. Although the rocker arms add mass to the valve train, they aid valve adjustment and servicing.

The rocker arms require guides that double as *buttons,* or *lash pads,* at each valve-stem tip. At the other end, adjustable pivot assemblies afford a convenient method of setting lash clearance.

Rocker arms are forged steel. To provide a durable rubbing surface where the cam lobe rotates against the rocker arm, a hardened steel *contact pad* is furnace-brazed to the top side of the rocker. Separate contact pads were not used at first. Instead, one-piece rocker arms were hard-chromed in the area of the cam-lobe rub surface. The one-piece rockers are no longer available.

The height of the rocker pivots is adjustable. This allows easy valve-lash adjustments. In 1983, the L28 280 ZX Turbo engine was equipped with hydraulic-pivot assemblies to reduce valve-train noise. These pivots, basically the same as conventional hydraulic lifters, don't move up and down with the cam lobe, but are

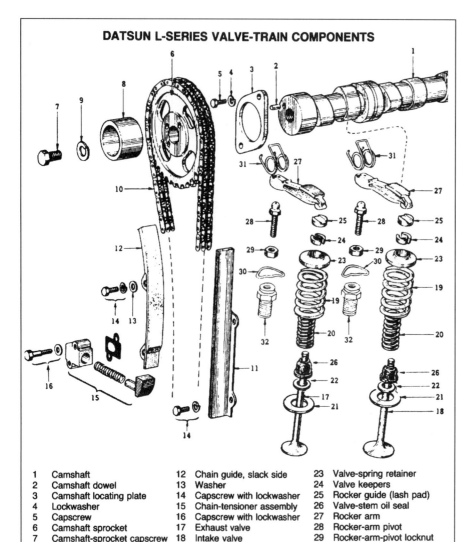

DATSUN L-SERIES VALVE-TRAIN COMPONENTS

1 Camshaft	12 Chain guide, slack side	23 Valve-spring retainer
2 Camshaft dowel	13 Washer	24 Valve keepers
3 Camshaft locating plate	14 Capscrew with lockwasher	25 Rocker guide (lash pad)
4 Lockwasher	15 Chain-tensioner assembly	26 Valve-stem oil seal
5 Capscrew	16 Capscrew with lockwasher	27 Rocker arm
6 Camshaft sprocket	17 Exhaust valve	28 Rocker-arm pivot
7 Camshaft-sprocket capscrew	18 Intake valve	29 Rocker-arm-pivot locknut
8 Fuel-pump eccentric	19 Outer valve spring	30 Rocker-arm spring retainer
9 Lockwasher	20 Inner valve spring	31 Rocker-arm spring
10 Timing chain	21 Outer valve spring seat	32 Rocker-arm-pivot bushing
11 Chain guide, tension side	22 Inner valve spring seat	Drawing courtesy Nissan.

stationary under the pivot in the cylinder head.

The lash pads, which are at the opposite end of the rocker arms, are installed between the rocker-arm tips and the valve-stem tips. Hardened lash pads reduce rocker-arm and valve-stem-tip wear by taking all the rubbing action from the rocker-arm tips and providing a larger contact area for the rocker tip. The stock lash pad is 3mm (0.118 in.) thick.

The stock valve-spring retainer is steel with a relatively short lash-pad support counterbored into the top surface. Its underside is machined to accept inner and outer valve springs. Its underside is stepped so the installed height of the inner spring is reduced to 1.378 in. compared to an outer-spring installed height of 1.575 in.

Stock valve springs have a combined seat *load*—frequently, but incorrectly referred to as *pressure*—of 74 lb. Stock L-series valve lift is 0.410 in. for all six-cylinder and L20B engines, and 0.390 in. for L16 and L18 engines. Stock valve springs approach *coil bind* at about 0.460-in. valve lift. A coil spring binds, or *goes solid*, when it is compressed so adjacent coils touch. At this point, the spring cannot be compressed any farther. Therefore, check valve springs for coil bind when modifying the valve train.

Valve-Train Lubrication—All L-series engines built after March, 1977 use *direct oiling*. The cams are drilled lengthwise and through each lobe for valve-train oiling. Oil flows up into the head, to the center two cam towers of fours or number-2 and -4 cam towers of sixes, into the cam, and sprays out at the lobes.

Four-cylinder L-series engines used direct valve-train oiling from their inception. Six-cylinder heads built prior to March, 1977 use a *spray bar* to lube the valve train. Oil flows up into the head, to the number-2 and -4 cam towers, into the spray bar, and out holes in the spray bar at the cam lobes.

For racing applications, spray-bar lubrication is superior to direct oiling. For hardware needed to make this conversion, see page 100.

STREET PERFORMANCE CAMS

The ideal street camshaft would provide significantly more power over a broader power band than the stock cam, but with no fuel-economy loss, rough idle or increase in exhaust emissions. Nice thought, but totally unfeasible. Most street-performance cams will improve power over a fairly broad band, but with fuel-economy loss and a moderate increase in emissions.

Different cam profiles or specifications will produce various engine-performance enhancements. The most desirable street-performance camshaft will increase power by 12—15% over the same power band as with the stock cam, but with a 3—5% loss in fuel economy and an insignificant increase in exhaust emissions.

Camshaft & Valve Train 63

L-series valve train consists of cam, rocker arms, lash pads, spring retainers, springs and valves. Valve lash is measured between cam-lobe heel and rocker-arm pad. Lash is adjusted at rocker-arm pivot.

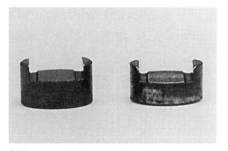

Different thickness lash pads allow adjustment of cam-lobe wipe pattern. Lash pads can be trimmed to exact size by using a lathe with a collet to hold the lash pads. Tom Monroe photo.

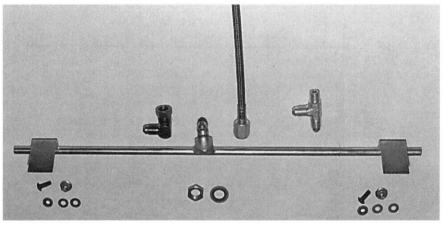

External spray-bar kit for four-cylinder engines is available from Nissan Motorsports. Spray bar bolts to valve cover, is fed by line routed from oil-pressure tap at side of block.

Semi-finished six-cylinder camshaft available from Nissan Motorsports, can be ground to your specifications. Cam is drilled for internal oiling so it can be use with later heads not equipped with spray bars.

The important thing to remember is that the stock L-series camshaft is the ideal compromise of all facets of engine performance. Any improvement in one area of performance generally results in a deficit in other areas. It isn't totally impractical for an avid performance enthusiast to choose a camshaft that will provide 20% more horsepower, but with the power band raised about 2000 rpm.

Such a cam will cause a lumpy idle and extremely high hydrocarbon emissions—excess unburned fuel will exit the combustion chamber. In addition to increased exhaust emissions, this type of street-performance cam will cause a 10—15% drop in fuel economy. Such a radical street cam will perform best with other engine modifications such as an exhaust header and a performance intake system.

Maximum Valve Lift & Lash-Pad Thickness—There are two factors to consider in street-performance-cam selection: maximum net valve lift and lash-pad thickness.

The maximum valve lift that can be used with stock L-series valve springs is 0.460 in. The maximum-thickness lash pad that can be supported in the stock spring retainer is 0.170 in. A cam profile with more lift or one ground so it requires thicker lash pads will require significantly more work to install the required components. For example, if you install an aftermarket cam with more than 0.460-in. net lift or a cam requiring lash pads thicker than 0.170 in., special valve springs and retainers will also be needed. This will add unnecessary expense and effort when

64 How to Modify Your NISSAN/DATSUN OHC Engine

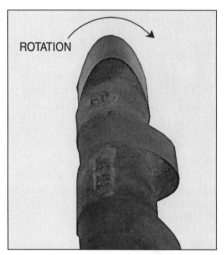

Symmetrical L-series cam profile provides good power with reasonable valve-train stability within limited rpm range.

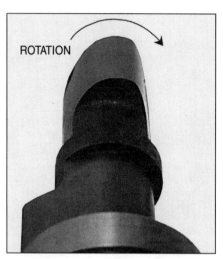

Asymmetrical cam-lobe profile provides good power and significantly improved valve-train stability. Engine rpm can be increased with less chance of component failure.

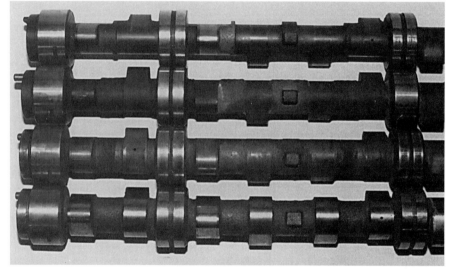

L-series cam lobes can be ground to virtually any profile. Bottom camshaft has stock grind. Next cam up has relatively mild lobe profile ground on stock cam. Third cam up has radical profile ground on a billet or unground stock camshaft. Cam at top has similar radical profile ground on stock cam. Note significantly different lobe base circles as represented by heels.

building a mild street engine.

Another factor that's not usually a problem with street-performance cams even on race cams with mild profiles, is *valve-train instability*. There are a few cam profiles that may cause problems at very high rpm, so be careful. For more on valve-train instability, read on.

RACE CAMSHAFTS

The main objective of using a race camshaft is to achieve maximum power over a desired rpm range and with assured reliability. These crucial requirements must be met by choosing the cam profile and compatible valve-train equipment. To select the proper cam, start by studying engine-performance needs, cam profile and associated valve-train components that will be used. For most road-race and drag-race applications, usable engine-speed range should be 5500—9000 rpm with peak power occurring in the 7500—8500-rpm area. By comparison, an off-road or rally engine should have a 3500—8000-rpm usable range and peak power around 5000—6000 rpm.

Valve-Train Instability—Under no circumstances should a cam or valve-train component be used that sacrifices reliability for power gain. Evidence of any of the following indicate that a cam profile or valve train is unstable:

- Broken valve springs.
- Significantly fatigued valve springs—loss of spring load.
- Displaced rocker arm.
- Broken valves.
- Significantly worn valve faces and seats.
- Broken oil spray-bar assembly.
- Severely worn timing chain, sprockets and guides.

One method of improving valve-train stability on high-rpm engines—those that operate at more than 8000 rpm—with radical cam profiles is to reduce valve-train mass. This can be done by installing lighter valves, springs and valve retainers. Lighter spring retainers are made from either aluminum or steel. A steel retainer that is properly designed and heat-treated for strength will be light, but very strong. Titanium retainers can be used, but, for the little bit of weight savings realized, they are very expensive and probably not cost-effective for most applications.

Valve-spring selection is also important because approximately half of the weight of a valve spring is part of valve-train inertia. It must control half of its own weight, too. Therefore, the best choice is the lightest weight spring with enough load to control valve-train mass. No more.

Another method of improving valve-train stability is to use an *asymmetrical cam-lobe profile*—cam-lobe *opening-ramp* profile is different than the *closing-ramp* profile. The profile is simply the shape of the cam lobe. An asymmetrical cam lobe rapidly opens the valve, but has a mild transition from the *nose* to the closing ramp. The closing ramp is contoured so the valve closes gradually, allowing the valve-train components to follow the cam lobe on the closing side. The valve closing action must be more gradual than on the opening side so the valve-train components stay in contact—*in mesh*—with each other.

Stable valve-opening and -closing action made possible with asymmetrical cam-lobe profiles will improve valve-train reliability and engine performance. The valves won't pound the valve seats or bounce off them when they should be closed and sealing the combustion chamber.

To make specific recommendations as to the cam you should use is not possible because each engine and application requires a different cam. And, more importantly, cam development is always ongoing. Chances are what was "the right stuff" yesterday won't be tomorrow, particularly in racing. When you're ready to choose a cam, go to an established source such as Nissan Motorsports, or a reputable cam grinder with experience supplying cams for Nissan/Datsun racing engines.

As a guideline, a "full-race" cam will have 0.580—0.630-in. lift, 310—320° duration, and be ground on 103—105° *lobe centers*—distance between each intake— and exhaust-valve cam lobe *at their points of full lift in crankshaft degrees*. An off-road or rally camshaft will have a 0.560—0.590-in. lift, 285—305° duration, and be ground on 98—102° lobe centers. An off-road/rally cam may also be suitable for oval-track racing.

RACE VALVE-TRAIN COMPONENTS

Valve-train components used for racing are similar in design to their stock counterparts. However, specifications are altered so these components are compatible with high-performance cam profiles and high-rpm engine operation. For instance, the valve springs must be adequate to control the valve train. And, the valve-spring retainers must match the dimensions of the springs and the required lash pads.

Rocker Arms—Stock rocker arms are suitable for all racing applications. As a testament to this, no aftermarket manufacturer I know of has bothered to make high-performance replacement rockers.

Regardless of their suitability, stock rocker arms should be inspected. Check the squareness of the cam-lobe rubbing pad on the rocker. This will affect cam-lobe wear and rocker-arm geometry. Also, check for an adequate amount of braze material holding the rubbing pad to the rocker.

Rocker-arm ratio is the numerical ratio of valve lift to lobe lift. Stock Nissan/Datsun rocker arms have an *approximate* ratio of 1.5:1, meaning the valve opens about 1.5 times the amount of cam-lobe lift. For example, if cam-lobe lift is 0.3167 in., valve lift is

1.5 × 0.3167 in. = 0.4750 in.

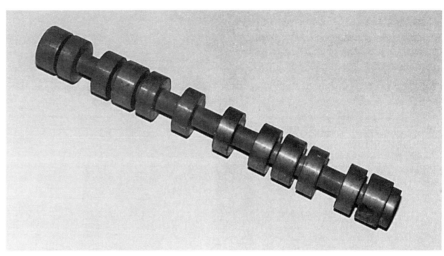

No, this is not a revolutionary camshaft. Billet camshaft has extremely "rough" lobes. Chances are cam grinder won't enjoy grinding cam.

I said *approximate* rocker-arm ratio is 1.5:1 because it changes slightly due to variations in contour and location of the rubbing pad on the rocker arm. Also, rocker-arm ratio can be changed slightly through the use of different-thickness lash pads and with readjusted valve lash. To check actual rocker-arm ratio, see page 79.

Rocker-arm weight is something else you need to check. First, the forgings should look the same. It's common for rocker-arm weights to vary significantly. The weight at the valve end of the rocker is what's important. After weighing and choosing a set of rocker arms, they can be lightened by grinding and polishing. The work should be done mostly at the valve end to reduce valve-train inertia.

Caution: Avoid installing used rocker arms, particularly if you're not going to install them with the same cam matched to the same lobes. Instead, install new rocker arms. Otherwise, you run the risk of destroying one or more cam lobes and rocker arms.

Valve Springs—Racing springs are

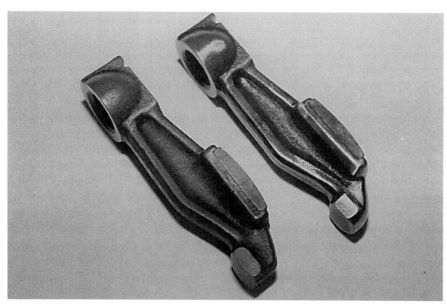

Stock rocker arm before and after prepping: Weight savings achieved at end of rocker reduce valve-train inertia—critical for high-revving race engine.

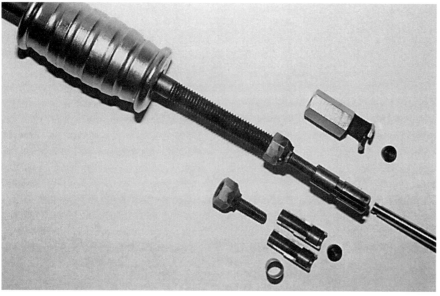

Electramotive used press-on steel wear caps with titanium valves. Wear caps must be removed from valve-stem tips with puller prior to valve-train disassembly. Top tool was modified by Don Reynolds to slip under cap. Tool kit at bottom was made by Electramotive. Either tool utilizes a conventional slide hammer as puller.

available in as many configurations as there are requirements. The ideal setup incorporates an asymmetrical cam for valve-train stability and lightweight valve-train components. This will allow a *light spring*—relatively low spring rate—with minimum *installed load*. Such a spring should stabilize the valve train, but not be overstressed.

Installed load occurs when the spring is at a height it would assume with the valve closed, or on its seat.

Depending on the cam profile and valve-train mass, L-series race springs typically have installed seat loads of 90—150 lb. A "stable" cam profile combined with titanium valves and other lightweight components may need no more than 90 lb on the seat. With a 200-lb/in.-rate spring, this type of seat load translates to about 230-lb spring load at 0.600-in. valve lift.

By comparison, a radical cam profile with steel valves, high-rate springs, and heavy retainers may require seat loads of 150 lb to provide valve-train control. This typically translates to a 350-lb-or higher load at 0.600-in. valve lift. Such a load usually results in valve-spring fatigue.

Nissan Motorsports offers a racing valve spring that works in most situations. By varying installed height, spring load can be varied over a broad range to match valve-train requirements. When used with steel valves and a reasonable cam profile, this spring is adequate up to engine speeds of 8000 rpm. For example, it can be installed at a height of 1.680 in. at 130 lb to accommodate a valve lift of 0.600 in. This setup will not overstress or fatigue the spring.

Electramotive offered a complete racing valve-train package: asymmetrical-profile camshaft, lightweight springs, matching retainers, valve-spring shims, titanium valves with steel wear caps, counterbored lash pads, and detailed installation instructions.

In this package, steel wear caps are pressed onto titanium valve-stem tips. These provide a harder surface at the tips of "soft" titanium valves that can cope with high valve-train loads. An L-series engine with this cam and valve train will produce excellent power to over 9000 rpm, with virtually no durability problems.

So far, the cam-and-valve-train package has been all good news. Now for the bad news. It follows the rule that says, "Going fast costs money. How fast do you want to spend?" This extensively tested package costs over $1000 for four-cylinder engines and over $1500 for sixes.

Although Electramotive is out of business, similar valve-train packages are available from Malvern Racing and Rebello Racing.

If you use a more radical cam and steel valves that could cause valve-train instability, consider using a "heavy" spring for

Stock spring seats shown here are not suitable for use with racing valve springs. Step or inner seat must be machined flush with outer seat to provide additional inner valve-spring installed height.

Cutter used to machine inner spring seat step flush with outer seat is powered with 1/2-in. electric drill. A photo of modified spring seats appears on the next page.

valve-train control. If sufficient spring load is not applied, valve float and/or valve-train breakage may occur. A properly selected heavy-duty spring installed at the correct height should be adequate up to the desired engine speed, providing this speed is "reasonable."

Such a condition will probably require a spring with a 150-lb maximum seat load and 350-lb open load at 0.580—0.600-in. valve lift. Springs such as this are available from various cam manufacturers. Although they may be intended for other applications, it's possible to apply them to your engine, providing you know the exact spring needed. Spring specifications you'll need include installed height and load valve lift, open load and coil diameters for the inner *and* outer springs.

Valve-Spring Retainers—Valve-spring retainers used on Nissan/Datsun L-series engines perform two functions. One is the standard two-part function of holding the spring in position relative to the valve while it closes the valve by transferring spring force to the valve stem. The second function is to contain and support the lash pad as it sits on top of the valve tip. Both of these functions and valve-spring and lash-pad dimensions must be considered when selecting retainers.

The underside of the retainer must fit the inner- and outer-spring coil diameters. Also, the retainer must align the springs so they remain concentric with the valve stem and combine with the spring seat at the head to give the correct installed spring height. The position of the underside of the retainer relative to the keeper groove affects installed spring height. In addition, the tapered bore at the center of the retainer must match the taper and OD of the *keepers*. Any change here will also affect installed spring height.

The lash-pad-support area at the topside of the retainer must be compatible with the thickness of the lash pad used. The lash pad should fit flush to 0.060 in. above the top of the retainer. *It must never be below the top of the retainer.* Nissan Motorsports' retainers are available with only one underside configuration. Differences in installed spring heights must be allowed for at the spring seat. However, three topside lash-pad-support heights are offered. These will accept three ranges of lashpad thicknesses: up to 0.180-in., 0.180—0.240-in. and 0.250-in.-and-thicker lash pads.

Special L-series valve-spring retainers are available from Nissan Motorsports in both steel and aluminum. The steel retainers are only slightly heavier than the aluminum ones because of the thinner cross section of the steel versions. Steel retainers have proven to be bulletproof. Although rare, a few aluminum retainers have failed.

Most aluminum-retainer failures can be traced to a problem that wouldn't affect a steel retainer. However, if the setup is correct, aluminum retainers will be adequate. In terms of cost, aluminum retainers are about half the price of steel retainers.

You can have titanium retainers custom-made, but for the slight weight reduction achieved, they are not cost-effective.

Keepers—Stock Nissan L-series retainer keepers are more than adequate for any high-performance application. I don't know of a single stock-keeper failure. The only reason not to use stock Nissan keepers for high-performance use is to substitute them with keepers that change the position of the spring retainer relative to the keeper groove.

Small-block Ford V8 keepers have a lock tang that is centered in each keeper—from top to bottom. The lock tang of L-series Nissan keepers is at the top. When Ford keepers are used in place of Nissan keepers, the spring retainer is raised about 0.050 in. in relation to the valve-stem keeper groove. All things being the same, this allows 0.050-in. more installed valve-spring height.

Special valve-spring retainers are sold by Nissan Motorsports. Three at left are aluminum; three at right are steel. Lash-pad support area of each is a different height to match to thickness of lash pad. Thicker lash pads require higher lash-pad supports and vice versa.

Modified spring seat has shoulder at inside and outside to register hardened-steel shim and, if required, selective-thickness shims.

If used, the top of the Ford keepers should not extend above the valve-stem tip. Otherwise, the lash pad will seat against the keepers instead of the valve-stem tip, resulting in the opening load of the valve being placed on the keepers and keeper groove rather than the valve-stem tip. Placing this kind of load on the keepers may cause them to unseat, resulting in the valve dropping into the cylinder. Piston destruction and severe cylinder-head and bore damage occurs shortly thereafter. To avoid such a problem, trim the top of the keepers so they'll be slightly below the valve-stem tip when installed.

Shims & the Spring Base—Valve-spring-base shims protect the aluminum spring seats from fretting caused by the motion and load of the valve springs. Shims can also be used to adjust valve-spring installed heights. Stock shims are not suitable for a racing setup because the offset, or *step*—raised area for the inner spring—does not allow the use of selective-thickness shims.

To correct this problem, the step, or inner spring seat, can be eliminated by machining it flush with the outer spring seat using a special cutter. At the same time, the inner/outer-spring-base location is lowered to increase installed spring height. The spring base can be lowered as much as 0.100 in. (2.5mm) without causing problems. However, cutting any more than this risks intersecting the intake-port roofs.

To determine the amount of machining required, you must know the recommended spring installed height for the application. The difference between this and the available installed height is the amount of material that must be removed from the spring base.

Spring-seat cutters are available from most high-performance camshaft manufacturers. For one, Iskenderian supplies a variety of cutters with different ODs and IDs and a selection of matching shims. These cutters are inexpensive and easy to use. Most can be powered with a 1/2-in. electric drill motor. Iskenderian shims come in 0.010-, 0.020- and 0.030-in. thicknesses. Whatever shims you use, make sure they are hardened. This will prevent fatigue resulting from the pounding of the valve springs.

Some applications require a thicker, smaller-diameter shim to adjust inner-spring load and installed height separately from the outer spring. Electramotive offered a complete package that does this. It includes the required base shims and installation and checking procedures. This package requires the use of an Iskenderian cutter that matches the Electramotive base shims and a thicker inner spring shim to match their spring setup.

Valve-Stem Seal—A good valve-stem seal to use on an L-series engine is from the Nissan/Datsun A-series engine. The seal from the A13 or A14 pushrod engine has a much smaller outside diameter, thus provides more clearance to the typically smaller ID of a high-performance inner valve spring. This is part 13207-H7210.

Because of increased valve lift, check

These are not new lightweight competition valve springs. Rather, they are set-up springs used to check installed height, valve lift retainer to seal clearance and piston-to-valve clearance.

Rocker-arm-tip and lash-pad wear patterns are also centered. This is ideal, but not absolutely necessary. Tom Monroe photo.

It's unlikely you'll need such a large selection of lash pads for setting up your engine. But installing a set of lash pads without checking and correcting cam-lobe wipe patterns by using different-thickness lash pads is a mistake.

for interference between the retainer or keeper underside and the top of the valve-stem seal. This can be done by trial-installing a valve with a lightweight spring and the correct retainer, keepers and seal.

Once installed, open the valve the same amount it will be opened by the cam. Check that retainer/keepers-to-seal clearance is at least 0.120 in. at full lift. If not, your alternatives include installing a shorter aftermarket seal for 0.312 in. or 8mm valve stems, shortening the valve guide or machining the underside of the retainer and/or keepers. Iskenderian makes a valve-guide cutter and offers special valve seals to work with shortened guides. Machine the retainer or keepers only as a last resort.

Lash Pad—Probably the most misunderstood valve-train component is the lash

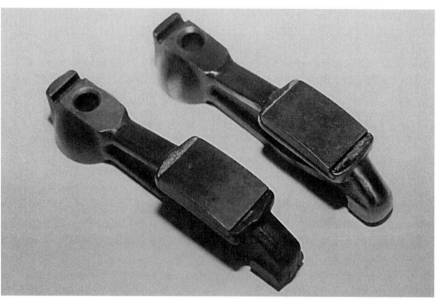

Lower rocker arm is unmodified. Race-prepped rocker arm at top was inspected, rocker ratio checked and polished for weight reduction. Contact pads are in alignment with both rocker arms. Start rocker-arm inspection by checking this alignment.

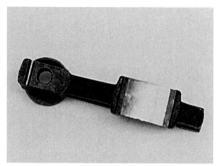

Cam-lobe wipe pattern on contact pad usually should be centered as shown. If additional lift is desired, pattern can be "cheated" toward rocker-pivot end and vice-versa. Wipe pattern must not go off contact pad. Layout bluing is used to check wipe pattern during engine mock-up, page 80. Tom Monroe photo.

pad, which doubles as the rocker guide. As explained earlier, the lash pad provides a large rubbing surface for the rocker-arm tip. Most L-series race cams are either *regrinds*—reground existing cams-or ground on standard cam cores. When increasing the cam lift, material must be removed from the base circle. Otherwise, the cam would not fit through the tamer bearing.

The resulting cam lobe base circle center line are shifted *away* from the rocker-arm rubbing pad. To compensate for this cam-lobe shift, the rocker arms must be moved parallel to its original position and up in relation to the cam lobe to maintain the correct rocker-arm geometry.

This is done with thicker lash pads. These lash pads, available from Nissan Motorsports, come in 0.010-in. increments from 0.150—0.330 in. thick. Stock lash-pad thickness is 3mm (0.118 in.). Rockerarm geometry is critical on the L-series engine if the cam is to survive. If it is not correct, the valves are side-loaded and this wears out the guides.

There is no predetermined thickness lash pad to use with a reground cam because many variables affect the thickness needed. The only sure way to determine the correct lash-pad thickness is to check the cam-lobe wipe pattern on the rocker-arm rubbing pad. The location of the wipe pattern on the rubbing surface can be moved by changing lash-pad thickness. The ideal wipe pattern is centered on the rubbing pad.

The adjustment of the wipe pattern on the rocker arm is critical with some race cams because the wipe pattern can be nearly as long as the contact pad. Therefore, it *must* be centered. Otherwise, the wipe pattern will go off one end of the rubbing pad, resulting in cam-lobe and rocker-arm destruction.

To increase rocker-arm ratio, the wipe pattern can be "cheated" closer to the pivot end of the contact pad. This is a mild "stroke," but if the wipe pattern is to be moved off-center, it should be moved in this direction because the lobe is in the closing mode and will wipe off the end of the contact pad. This is safer than allowing the opening ramp of the lobe to contact the leading edge of the contact pad.

For checking the lash-pad thickness and wipe pattern, see Chapter 7, pages 79-80.

CHAPTER SEVEN
Engine Preassembly

John Teaby's IEM Racing PL510 won the 1999 Nasport GT4 Championship with Rebello Racing L16 engines. G. Hewitt photo.

Components and tolerances of a racing engine are significantly different than Nissan/Datsun stock specifications. Consequently, engine *preassembly* is mandatory to ensure that correct clearances are achieved. During this process the correct fit and adjustment of *all* engine components is done. This is also an excellent time to double-check all of your prep work and modifications.

CYLINDER BLOCK
Start with the cylinder block by finding the pipe plugs that install in the main oil gallery. Remember, the plug that goes in the front of the main oil gallery must be shortened at the outer end so it installs flush or slightly below the front face of the block. Shine an inspection light into the number-1 main-bearing oil passage and peer down it to check that the other end of the plug doesn't protrude so far into the gallery that it blocks off the passage. If it does, remove the plug and shorten the inner end.

Install the other plug in the rear of the main oil gallery and check its fit. It should be flush or below the rear face of the block. Don't worry about the inner end. When all is OK with the two plugs, remove them and set aside for final assembly.

Cut-away engine would help engine assembler check component fit visually and would provide access for measuring tools. However, on actual engine, you must use other measuring methods and a trial assembly to ensure engine components fit correctly prior to final assembly.

PISTONS & RINGS

Check Pistons & Bores—Using clearance and measuring recommendations from the manufacturer, check the bore clearance of each piston. Try them in different bores.

The best way to check piston-to-bore clearance is to measure the piston-skirt diameter with a micrometer at the location recommended by the manufacturer. After you've measured piston diameter and recorded the results, use the micrometer set to this dimension. Now, lock the micrometer and use it to zero a dial bore gage. The dial bore gage can now be used to measure the bores to determine how much larger each is than the piston. Once you've matched a piston to each bore, scribe the bore number on the underside of the pin boss. For higher visibility or temporary marking, use a felt-tip marker on the piston top.

While measuring the bores, check them for *out-of-round* and *taper*. After zeroing it in the bore, rotate the gage in the bore to check for out-of-round. Out-of-round should be less than 0.001 in. and definitely not over 0.002 in. Again, zero the gage and move it down from the top of the bore to check taper. As with out-of-round, taper must not exceed 0.002 in.

Insert the rings into the piston and check backside and side clearance. Backside clearance is the distance between the back of the ring and the ring groove. To check backside clearance, insert the edge of the ring in each groove and measure how far it is below the outside surface of the piston. Ideal backside clearance is 0.005—0.020 in. Up to 0.030-in. backside clearance is OK.

Side clearance is the vertical distance between a ring and its lands. Check side clearance by installing the rings and measuring between the rings and lands with a feeler gage. Acceptable ring side clearance is 0.0015—0.0035 in.

Next, each set of rings should be checked in *its* bore, then kept in order. This is because the rings should be installed in the bore they were fitted to.

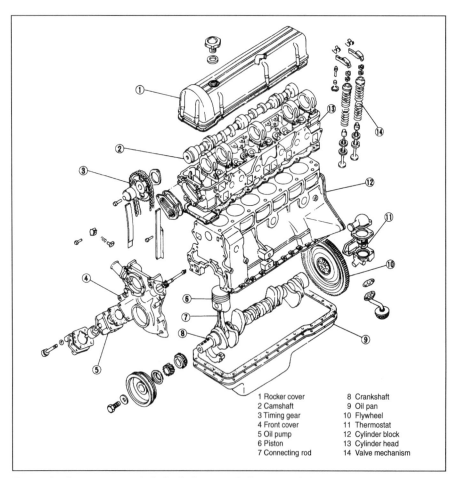

1 Rocker cover
2 Camshaft
3 Timing gear
4 Front cover
5 Oil pump
6 Piston
7 Connecting rod
8 Crankshaft
9 Oil pan
10 Flywheel
11 Thermostat
12 Cylinder block
13 Cylinder head
14 Valve mechanism

Except for thermostat, crankshaft oil slinger, and oil pump and pickup, race engine has same basic components as stock engine. Drawing courtesy Nissan.

Engine Preassembly

Dial bore gage and 3-4 in. micrometer are best tools for checking bore taper and out-of-round. Here, dial bore gage is used to check bore during piston-fitting process. Tom Monroe photo.

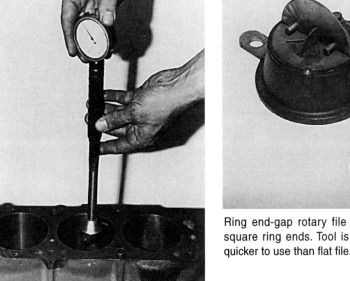

Ring end-gap rotary file helps to ensure square ring ends. Tool is also easier and quicker to use than flat file.

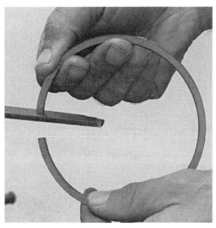

If using flat file to gap rings, clamp file in vise and move ring against file in direction shown. Afterwards, deburr corners of filed ring end. Tom Monroe photo.

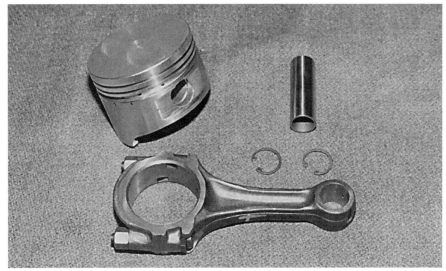

Piston, pin and retainer assembly checked for pin end float and ready for installation into engine. Pin was too long as it extended past retainer groove. To achieve correct fit, pin ends were lapped on sheet of sandpaper to shorten it about 0.004 in.

Ring End Gap—To check piston-ring end gap correctly, the ring must be fitted in the cylinder bore and squared. To square the ring, use some caution and a flat-top piston. With the piston held upside down, push the ring partway down in the bore. Check end gap with your feeler gages.

Piston-ring end-gap specifications vary according to certain factors, bore size being the major factor. If the manufacturer didn't supply ring-gap specs, a good general rule to follow for compression rings is to allow 0.0035—0.0040 in. end gap for every inch of bore diameter. For example, a 3.400-in. bore will require a ring end gap of $0.0035 \times 3.4 = 0.0119$ in. at the low end and $0.0040 \times 3.4 = 0.0136$ in. at the high end. Some engine builders add 0.002 in. end gap to the top ring on an engine that produces relatively high power. This is to cope with high combustion temperatures. Such temperatures require more gap to allow for increased ring expansion.

To increase ring end gap, use a special ring-end rotary file or a flat file held in a vise. Always remove equal amounts of material from each end of the ring and keep the ends square and parallel. On rings with coated faces such as chrome or moly, use caution not to chip the coating. Do this by using a very fine flat file and stroking the ring end against the file from the outside edge of the ring to the inside edge. The edges of the ends should be lightly sanded to deburr the corners once the rings have been fitted.

PISTON PINS

Check the fit of each pin in its piston. A piston-to-pin clearance of 0.0008—0.0011 in. is desirable. If the clearance is correct, the pin should be tight in its bore, but should slide fairly easily through a clean, dry pin bore. The feel should be the same as it should in the small end of the rod.

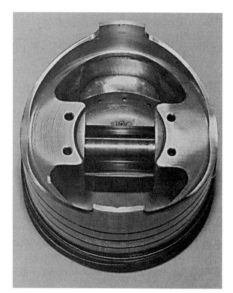

Although pin was run over 60 hours under race conditions, pin and bushing show minimum wear. Initial pin-to-bushing clearance was 0.001 in. Note that hard edges on piston were polished to break them.

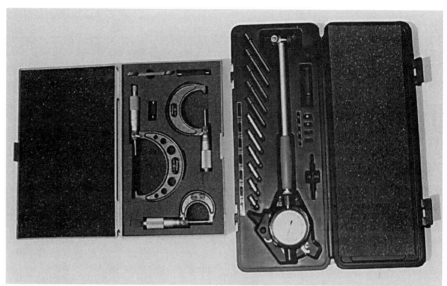

Micrometers and dial bore gage are ideal for checking cylinder- and bearing-bore sizes and clearances.

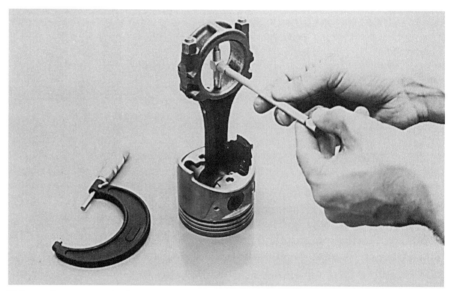

Telescoping gage and micrometer can also be used to check connecting-rod big end. Tom Monroe photo.

If you are using full-floating piston pins with Teflon buttons at the ends, measure the overall length of the pin-and-button assembly. This measurement should be 0.015—0.017-in. less than the bore diameter to provide sufficient pin-to-bore clearance. If the pin-and-button assembly is too short, get new buttons. But if the assembly is slightly too long—about 0.002 in. you can machine the buttons to make them shorter, providing you maintain the original contour. Pin-and-button assemblies that are substantially longer should be corrected with new buttons.

If beveled-type circlips or Spirolox-type retaining rings are used, install a ring in one groove only in each piston-pin bore. Install the circlip so the rounded edge will be against the pin. This will put the sharp, square edge out, away from the direction of pin thrust.

Insert the pin in the other end of the pin bore, hold it against the retaining ring, and check the distance from the end of the pin to the edge of the empty retaining-ring groove. A 0—0.001-in. clearance is desirable. This will keep the pin from being preloaded against the retainers and possibly popping them out of their grooves.

After checking and recording the results, remove the piston pin. Leave the retainer in place. Keep each pin with its matching piston as a set for final assembly. Next, check the fit of the pin in the small end of the respective connecting rod. Pin-to-bore clearance should be 0.0008—0.0011 in.

Caution: Never reuse Spirolox type retaining rings. Instead, replace them.

CONNECTING-ROD BEARINGS

To select the correct rod bearings, start by checking the size of the big end of each rod. If the bore in which a bearing is to be installed isn't the correct size or true, bearing fit or crush will not be correct. So, if you haven't checked the big ends for size and trueness, do so now.

Install the cap on each rod as it would

Engine Preassembly

Sunnen® gage us used to check connecting-rod big ends during reconditioning process. Rod is rotated against flat plate and dial indicates out-of-round similar to dial bore gage. Tom Monroe photo.

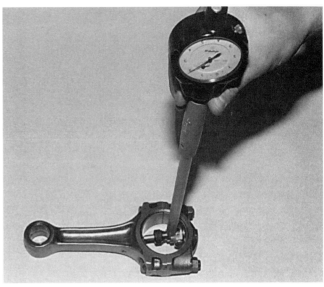

Dial bore gage is used to measure connecting-rod bearing-bore ID to determine journal-to-bearing clearance. Crank-pin diameter is measured with outside micrometer.

Rear thrust face of main bearing is damaged. Probable cause was inadequate crankshaft end play or excessive clutch pressure-plate load. Note that oil-supply hole was enlarged.

be in the engine-cap and rod numbers matched and aligned. With the rod positioned upside down, clamp the big end of the rod in a vise with soft jaws, then torque the nuts or bolts to spec. Hold the big end in a vise while torquing the nuts. If stock Nissan 9mm bolts are used, torque the nuts 33 ft-lb (4.5 kg-m). If aftermarket bolts or special rods are used, torque the nuts or bolts to the low end of the manufacturer's specification.

Use a micrometer to zero the dial bore gage at the specified big-end size. Then measure each rod big end in several locations—across the parting line and in line with the beam. If the connecting rods aren't to specification, have them resized.

To select bearings and check bearing clearances, measure each crankshaft connecting-rod journal in four locations with a 1-2 in. micrometer. Record each measurement. Select the bearing shells that should give 0.0023—0.0028 in.-bearing-to-journal clearance and install them into a connecting rod.

Assemble the rod with the bearings in place and torque the nuts or bolts to spec.

Remove the rod from the vise and, with the dial bore gage adjusted to the respective *journal size,* carefully measure rod-bearing ID. The reading you get is bearing-to-journal clearance. Disassemble the rod and mark the backside of the bearing shell with a permanent felt-tip marker or a fine-tip scribe to indicate the rod number. Also indicate U or L for upper or lower shell, respectively.

If bearing clearance isn't correct, repeat the assembly-and-checking process with a different-thickness bearing shell(s). Note: It's OK to have up to a 0.0005-in. variation in thickness between upper and lower shells to achieve the desired bearing clearance.

MAIN BEARINGS

Once you've finished selecting rod bearings, turn your attention to the mains. Use the same checking procedure for selecting main bearings as you used for the rod bearings. First, install a bearing shell in the block and cap bores. Install the cap. Torque standard main-cap bolts 51—61 ft-lb (7.0—8.5 kg-m); diesel bolts 58—65 ft-lb (8.0—9.0 kg-m).

With a 2-3 in. micrometer, measure the main-bearing journal. Adjust the dial bore gage with the micrometer set to journal diameter. Measure bearing ID in three places-vertical, and at 45° to vertical on each side. Exchange the bearing shells as needed to achieve 0.0022—0.0027 in. clearance. As with rod bearings, different-thickness shells can be used to achieve the desired main-bearing clearance.

Once the correct clearance is achieved, remove the shells and mark them as to journal number and B for block or C for cap. Do this for each journal.

If you're wondering why I haven't mentioned using Plastigage® for checking clearances, I prefer using mikes and

Crankshaft end play is checked with dial indicator set square against the nose. Crank is forced forward in block, indicator is zeroed as shown, then crank is forced rearward to give end play, or float. Tom Monroe photo.

Prior to short-block mock-up, final check of crankshaft straightness is made. Although crank is extremely tough, moderate impact on crank can cause an out-of-round condition!

gages for maximum accuracy. I don't trust the measurements from using Plastigage.

Check Crankshaft End Play— After you've selected all inserts, check crankshaft end play, or thrust. Install the center main-bearing shell and any other two shells in the block and the other halves in their respective caps. It is not necessary to oil the bearings. Just don't turn the crank once it's installed. Install the crankshaft, then the bearing caps. Torque the respective cap bolts to spec.

Attach a dial indicator with a magnetic base to the rear of the block. Align the plunger so it is square to the rear face of the flywheel flange. Force the crankshaft forward in the block, zero the dial indicator, then force the crank rearward. Read end play directly from the dial indicator.

End play should be 0.0025—0.0040 in. If it is less than 0.0025 in., you can increase end play by removing material from the bearing thrust faces. Do this by carefully polishing the thrust flanges of the center main bearing with 320-grit sandpaper on a flat surface. Before doing any polishing, measure the thickness of the bearing thrust flange with a 0—1-in. mike. Measure the mating flange of both shells at both ends and in the middle. Record the results.

To polish, hold the shells together as they would be installed in the block and face down on the sandpaper. Remove only the material that's needed to obtain minimum end play. Work the shells back and forth with one face down. Stop frequently and check your progress so you don't remove too much material.

Enlarge Oil Holes—If your engine is a four-cylinder and the number -2 and -4 main-bearing oil passages have been enlarged, the holes in the respective upper main shells will also have to be enlarged. They must also be matched to the holes in the block. With each bearing shell clamped in a vise and, using a drill press, enlarge the hole with a 5/16-in. drill. Deburr the edge on both sides of each hole with a *single-flute*—one cutting edge—countersink. Use a sharp, fine-tip knife or, if available, a double-edged, dental-gold knife to finish chamfering the hole around the bearing oil groove.

Burrs at edges of threaded holes (shown at left) should be removed. Photo at right is of same holes after deburring—chamfering, in this case. Chamfering threaded holes reduces chance of pulled threads.

SHORT-BLOCK MOCK-UP

During connecting-rod preparation, the center-to-center distance of each rod was checked. And, during piston prep, the compression height of each piston was checked. Consequently, if you install one piston-and-rod assembly into the block and onto the crank, it should reflect the fit of the remaining piston-and-rod assemblies. Therefore, you need only to check one piston-and-rod assembly.

To do your mock-up assembly, install the crankshaft with two not-so-worn used bearings on at least two journals. The front and rear journals are good choices. Used bearings are OK on the rod, too. Install the number 1 piston-and-rod assembly with the top ring only—with a used ring, if one is available.

Slide the crank sprocket into place, then install a large-diameter degree wheel on the front of the crank. The crank-damper/pulley bolt should fit the degree-wheel center hole.

Make a pointer from a piece of heavy wire—gas-welding rod will do. You'll need about a 6-in.-long piece of wire. Grind one end to a point and bend about a 1/4-in.-diameter loop in the other end. Secure the looped end to the front of the block at a front-cover bolt hole. Bend the wire so the pointer tip is directed at the edge of the degree wheel and parallel to its plane. Rotate the crankshaft so the number-1 piston is at or near TDC. Position the degree wheel so the tip of the pointer is near TC on the wheel. Later, you can bend the

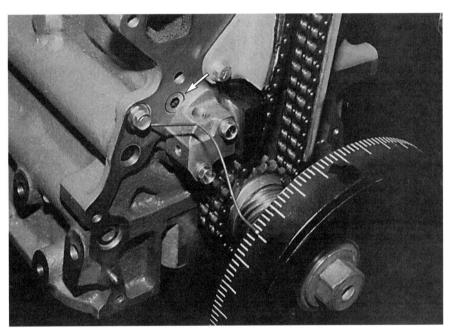

Note oil-gallery plugs in front of block. Top plug is installed as it must be—flush with front of block (arrow). Plug seals main oil gallery. Remaining two plugs seal oil-pump-supply and discharge galleries because engine will use dry-sump oil system. Consequently, these passages must be plugged.

pointer tip to make the final adjustment. Keep the crank from rotating and moderately tighten the damper/pulley bolt to secure the degree wheel. Attach a magnetic-base dial indicator to the top of the block and align the plunger vertically so it's square to a flat area on the piston top.

To determine exact TDC, rotate the crank back and forth while the piston is close to TDC and observe the indicator. The crank is at TDC when the indicator reverses direction as you rotate the crank. Zero the dial when the piston is at the top of the stroke. Recheck TDC and adjust as necessary. Once you are sure the piston is at TDC, adjust the pointer to TDC on the degree wheel. From now on, be very careful not to move the pointer or degree wheel until you're ready to remove it.

Piston-to-Head Clearance—Check the distance between the flat portion of the piston top and the head. To determine

Six-cylinder race head has welded combustion chambers, and oversize intake and exhaust valves. Volume of each combustion chamber must be measured and compared. Depth of intake and exhaust valves must also be checked.

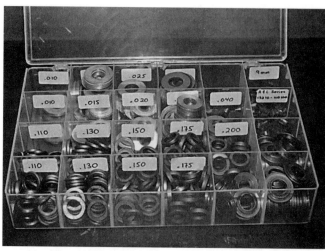
Variety of valve-spring base shims: Thick inner-spring shims adjust inner-spring installed height. Shims that install against spring must be hardened steel.

this clearance, one dimension you must know is the installed—crushed—thickness of the head gasket. Head-gasket thickness plus or minus piston *deck height*—distance of piston top is *below* or *above* the block deck—gives piston-to-head clearance. Minimum piston-to-head clearance is 0.050 in.

To determine piston-to-head clearance, you must first determine piston deck height. While the piston is at TDC and the indicator is on the piston and zeroed, lift the plunger and rotate the dial indicator from the piston to the block deck. While moving the indicator, be careful not to alter the position of the holder frame. Lower the plunger onto the deck and read deck clearance. If the piston is above the deck, treat this clearance as a negative value. Repeat the process to confirm accuracy. Remove the dial indicator and leave the engine assembly as is.

Cover up the block so it'll remain clean while you check the head and prepare it for trial installation. If the block is on a stand, slip a large garbage bag over it to keep it dust-free.

CYLINDER-HEAD ASSEMBLY

Prior to mocking up cylinder head, make sure you've recorded the following dimensions and/or performed the following operations:

- Determined combustion-chamber volume, measuring and matching all chambers.
- Set valves to equal depth in chambers.
- Modified valve-spring seats.
- Selected and specified valve-spring shims and retainers.
- Selected valve seals.
- Specified and selected camshaft.
- Specified cam-tower shims, if required.

Valve Springs—Because you've already selected the valve springs, you should have their installed heights and loads. You'll have to check the springs for load at open height. It depends on valve lift, which depends on cam-lobe lift. With this in mind, the springs should also be checked for coil bind at maximum valve lift, if you haven't already done so.

Some engine builders use less valve-spring load on the exhaust valves because they are lighter than the intakes. To achieve this with the same springs, simply use more installed height on the exhaust valves. Another method is to use the same type of springs on both valves, but with springs of a lower *rate*, or force per unit deflection—pounds per inch—on the exhaust valves. Whatever method you choose, you should have spring specs prior to installation.

First check the entire set of springs for quality. The ground ends should be flat and perpendicular to the spring center line. Check this with a square. The coils should be evenly spaced. All the springs should be checked for load on the seat and at full valve lift. A maximum spring-load variation of two percent is allowable.

To check the springs for coil bind, compress them to their height at full valve lift. You'll need a spring tester or vise and vernier calipers for this. There should be a minimum of 0.015 in. between the coils or about a total of 0.075 in. from spring height at full valve lift to the height at which the valve spring stacks solid. Check both inner and outer springs for coil bind.

To check for retainer/keeper-to-valve seal interference, install an intake valve and an exhaust valve. Use a lightweight spring in place of the valve spring. Assemble the lightweight spring with the retainers and keepers that will be used. Then, open each valve to full lift and check distance between the underside of the retainer/keeper to the top of the valve

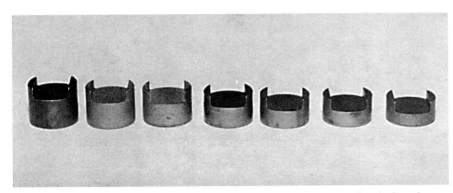

Lash pad at left is 0.330-in. thick; pad at right is 0.150- in. thick. Intermediate lash pads vary in thickness by 0.010 in. Some are shown.

Undersides of lash pads can be counterbored to allow for thickness of wear caps used with titanium valves. Wear caps are usually 0.100-in. thick as measured at their centers.

seal. As detailed in Chapter 5, minimum retainer/keepers-to-seal clearance is 0.120 in. at full lift.

Valve-Spring Installed Height— As you may remember from the valvetrain chapter, some valve springs will require that the valve-spring seat be machined. And most valve-spring combinations require the removal of the step for installing the inner spring. This is easily done by machining the spring seat with a cutting tool powered by a 1/2-in. electric drill, page 67.

There are several methods of measuring valve-spring installed height. What you must be able to do is accurately measure installed height and shim the spring as required. The most common measuring method is with a telescoping gage and vernier caliper or outside micrometer. While holding the valve, retainer and keepers with a lightweight coil spring, measure installed height with the telescoping gage and lock the gage in position. Then, measure the gage with a vernier caliper or outside mike.

Special-made micrometers for measuring installed height directly are also available. The special mike installs in place of the spring and is expanded until snug. Installed height is read directly off the barrel of the mike.

Perhaps the simplest tool for checking installed height is a piece of coat hanger or welding rod cut to the desired dimension.

Valve-spring shims are typically available in 0.005-, 0.010-, 0.015- and 0.020-in. increments. After selecting the correct shims, organize and store them so they won't be mixed up before or during assembly.

Check Rocker-Arm Ratio— To check rocker-arm ratio, place each rocker arm on the same cam lobe, lash pad and valve assembly. Set up a dial indicator with its plunger square against the spring retainer and in line with the valve stem. With the valve on its seat, zero the indicator. Rotate the cam and check maximum valve lift. Valve lift from one rocker arm to the next rocker should not vary by more than 0.005 in.

While you're at it, visually check the squareness of the cam-lobe rubbing pad on the rocker. If the pad is tilted, the load will be concentrated on a small part of the lobe and rocker pad, leading to accelerated wear. Also, check for the presence of adequate braze material on the pad. If there are any voids or gaps in the braze, the pad could break loose from the rocker and cause serious damage.

After checking the rockers, select a "set" to be final-weighed and balanced. When balancing the rocker arms, remember that the weight at the valve end of the rocker is what's important.

Because numerous rocker-arm forgings have been used in L-series engines, rocker-arm weight may vary considerably. Balance the weight of the arms to match the lightest one by grinding, as described in Chapter 6.

Lash-Pad Selection— Several factors determine what thickness valve-lash pads should be used. The major factor affecting required lash-pad thickness is created by the use of an aftermarket reground camshaft, where material is removed from the cam-lobe base circles. Because this requires raising the rocker arms at the pivot ends to maintain valve lash, the valve-tip ends must also be raised to maintain rocker-arm geometry. To accomplish this, thicker lash pads are installed.

Other factors that determine required lash-pad thickness are height of the valve-stem tip, the use of cam-tower shims and the removal of material from the top surface of the cylinder head. If a valve-stem tip is raised, a thinner lash pad is used; thicker, if the tip is lowered. The use of cam-tower shims or milling the top of the head have the same effect as a reground cam—thicker lash pads are used. Choosing lash pads is explained in Chapter 6.

Valve-lash pads, which install in the spring retainer at the valve-stem tip, are

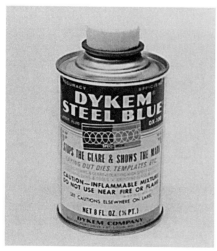

Layout blue is invaluable tool for use in building race engine, particularly when checking cam-lobe wipe pattern. Tom Monroe photo.

Bright link is aligned with crank-sprocket timing mark. Location of bright link (arrow) at cam sprocket is determined during cam timing. Tom Monroe photo.

Although not as important as cam-lobe wipe pattern, tip wipe pattern on wear cap should be near center. Unlike lobe wipe pattern, wear-cap pattern cannot be adjusted. Tom Monroe photo.

available from several sources. One source is Nissan Motorsports, where thicknesses range in increments of 0.010 in. from 0.150 to 0.330 in. Stock lash-pad thickness is 3mm (0.118 in.).

To determine the needed lash-pad thickness, perform the following check.

Cam-Lobe Wipe Pattern—If you're installing an aftermarket reground cam or you've shimmed the cam towers, milled the top of the head or ground valve faces or seats a considerable amount, you must check cam-lobe wipe pattern.

As mentioned earlier, the cam-lobe wipe pattern must not extend off either end of the rocker-arm pad. It should be centered on the pad. To adjust the wipe pattern, use different-thickness lash pads. To determine which thickness lash pad to use, perform the following trial installation.

Choose a lash pad and install it on the number-1 exhaust valve. Apply layout blue or permanent marker to a rocker-arm pad. Install the cam and rotate it so the number-1 exhaust-lobe toe is straight up, away from the head. Compress the valve spring and install the rocker arm on the pivot and valve tip. Adjust valve lash to the cam manufacturer's specs.

Wipe the rocker arm by rotating the camshaft one full revolution. The rubbing action of the cam lobe will remove bluing or marker from the rocker-arm-pad area contacted by the lobe.

Compress the valve spring and remove the rocker arm. Inspect the cam-lobe wipe pattern. It should be centered or almost centered on the pad. If the pattern is toward the valve end of the rocker, install a thinner lash pad, readjust lash and recheck the wipe pattern. If closer to the pivot end, try a thicker lash pad. If the pattern position is reasonably well-centered, remove the rocker arm and lash pad.

Continue selecting lash pads and checking the wipe patterns of each rocker arm until you've checked them all. Store and identify the lash pads and rocker arms so you can install them on the same valve at engine-assembly time.

Time Cam—Once you've finished selecting the lash pads, reinstall the number-1 intake- and exhaust-valve lash pads and rocker arms. Rotate the cam until both intake and exhaust valves are closed. This positions the cam near its TDC compression-stroke position for number-1 cylinder. The cam-sprocket dowel at the front of the cam should be straight up relative to the cylinder head.

INSTALL HEAD

Uncover the block and place a used head gasket on the deck surface. If you removed the two hollow head-alignment dowels from the block deck, reinstall them first. They install in countersinks at cylinder-head bolt holes on the manifold side of the block.

If you moved the crank, rotate it back

Engine Preassembly 81

Nissan Motorsports offers chain-tensioner assembly with sheet-metal limiter. Limiter prevents shoe from being pushed from tensioner body. Tom Monroe photo.

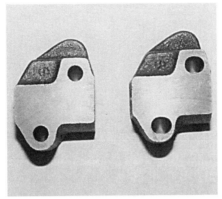

Stock timing-chain tensioner body is at left. Body at right has elongated bolt holes to allow slight inward adjustment of tensioner assembly. Curved chain guide should also have elongated holes if tensioner-body holes are elongated.

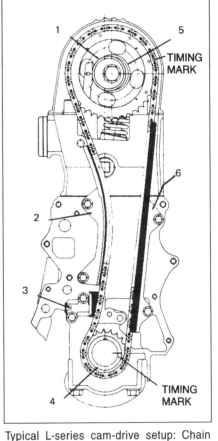

CHAIN INSTALLATION
1 Fuel pump eccentric
2 Chain guide—"slack" side
3 Chain tensioner
4 Crank sprocket
5 Cam sprocket
6 Chain guide—tension side

Typical L-series cam-drive setup: Chain should have smooth arc between tensioner shoe and bottom of curved chain guide. Drawing courtesy Nissan.

to the number-1 TDC position. Set the head on the block and secure it with two bolts. **Caution:** During this trial assembly, be careful not to bump the timing-wheel pointer.

Install Cam Drive—With the cam and crank set at their TDC positions, install the timing chain, sprockets, guides and tensioner. Use minimum Grade-9 bolts to secure the chain guides and tensioner. Install the cam sprocket in the best-guess position: 1, 2 or 3 hole at the cam sprocket.

Install the cam sprocket with the timing chain draped over it. The chain should be positioned so the two "silver," or bright links are at the front. The fewest number of links between the bright links will be on the distributor side of the engine when the chain is installed. Align one bright link to the crankshaft-sprocket dot, which should be at about the 4 o'clock position when installed.

Align the second bright link to the selected cam sprocket—start with dot 1. The link/dot position on the cam sprocket should be at approximately 2 o'clock, with the corresponding dowel hole in sprocket hub aligned to the cam dowel. Install the cam sprocket to the cam nose. Double-check that cylinder 1 is at TDC and check the cam-sprocket notch-to-dash-mark position.

Alignment of the *slack-side* guide—the curved one—to the tensioner is crucial.

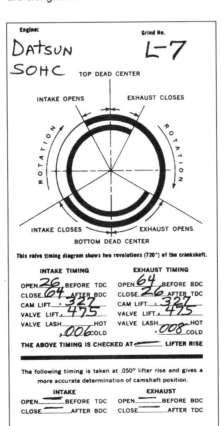

Cam-spec card from Iskenderian Racing Cams is similar to those from other cam manufacturers. Refer to specifications that accompany cam when checking its timing.

The timing chain should make a continuous arc across the face of the guide and the tensioner shoe while the tensioner shoe is fully retracted in the tensioner body. In this position, the timing chain should contact the center of the tensioner shoe, not the top or bottom edges. To achieve this, you may have to elongate the already elongated bolt holes in the slack-side guide or tensioner body. Also, the tensioner can be rotated slightly after loosening its bolts to align the shoe face with the chain.

To check camshaft timing correctly,

With cylinder head off, cylinder-1 TDC is easily found with dial indicator set square against top of piston. Crank is rotated back and forth until maximum reading is found to establish TDC. Tom Monroe photo.

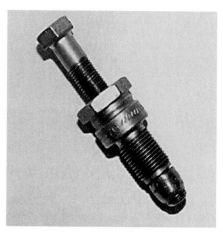

Modified spark plug is used to find cylinder-1 TDC with cylinder head installed. Tool is inserted into spark plug hole and adjusted to stop piston travel at about 10 before and after TDC. Crank pointer is then adjusted to align with TDC on degree wheel.

you must have a cam-specification card. This card, which should've been supplied by the cam manufacturer, usually contains all information, such as valve lift, cam lift, intake and exhaust events and valve lash, you'll need for checking your cam. Most cam grinders measure valve events beginning and ending at different lifts, so depend on the information on the spec card when checking your cam.

Check Valve Timing—Utilizing a short bolt in a valve-cover bolt hole, mount a steel plate to the cylinder-head valve-cover surface. It doesn't have to be very big. You now have something to mount your magnetic-base dial indicator to.

Align the dial-indicator tip vertically on the flat surface of the number-1 exhaust-valve retainer. The exhaust valve should be closed if you haven't turned the crank. Zero the dial.

To make it easier to compare the manufacturer's cam specifications to those you'll obtain from the following procedure, copy the same basic layout of the manufacturer's card on a separate sheet of paper. You can then use this sheet for recording and comparing numbers.

Rotate the crankshaft until the exhaust valve begins to open and stop when the specified *opening* valve-lift number is indicated on the indicator dial. Record the position of the crankshaft indicated by the pointer at the degree wheel. Rotate the crankshaft until the exhaust valve is about closed, but stop at the specified *closing* valve-lift number. Record the position of the crankshaft indicated at the degree wheel. Perform the same process on the intake valve.

Compare the recorded valve-event numbers to those specified by the cam card. Some cam grinders will even provide a valve-lift figure at TDC valve overlap for both intake and exhaust valves. A valve-lift check is relatively easy to make at TDC valve overlap and is helpful in determining cam timing. For example, if the intake-lift dimension is higher than spec and exhaust-valve lift is lower than spec, the cam is advanced. Conversely, less intake lift and more exhaust lift indicates retarded cam timing.

Because cam timing is measured as lift or events at the valves, rocker-arm ratio becomes a significant contributing factor. Setting cam timing to spec confirms that valve operations occur at the correct time in relation to piston locations.

It is possible to have the cam correctly timed, but have the recorded numbers vary slightly from the cam manufacturer's specs. If rocker-arm ratio varies from stock, duration will be affected. For example, a greater rocker-arm ratio allows more duration. To maintain cam-timing lobe-center positioning to specs, compensate for the high rocker-arm ratio by opening the valve 1—2° sooner and closing it the same amount later, thus splitting the duration.

If cam timing is incorrect, reposition the cam sprocket on the cam. The stock sprocket has three positions: 1, 2 and 3. The difference between each position is 4° of crankshaft rotation. For applications requiring a wider range of sprocket adjustment, Nissan Motorsports offers an eight-hole sprocket. Each of these positions provides 2° of crankshaft rotation. Another

Nissan Motorsports eight-hole (eight position) cam sprocket replaces stock three-position cam sprocket. Holes are spaced so cam timing can be changed in 2° *crankshaft* increments.

Offset bushings are inexpensive way to adjust cam timing. Sprocket hole is drilled oversize to 3/8 in. to match bushing OD.

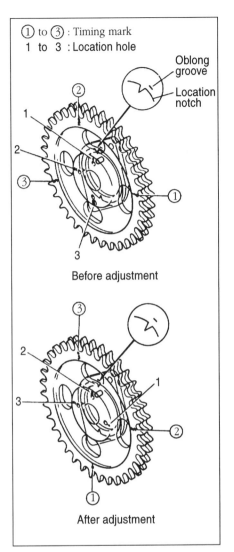

If cam-sprocket notch aligns with thrust-plate groove with crank set at TDC, cam is timed. If notch is to the left of groove (left), cam is retarded. Remove and reposition sprocket clockwise to next dowel hole and align 2 timing mark at teeth—if originally at 1—to bright link. Recheck notch alignment with groove. Move to 3 groove and mark if originally set at 2. Drawing courtesy Nissan.

way to get more adjustment is to use offset bushings in a sprocket with enlarged dowel holes.

Check Piston-to-Valve Clearance— When you've finished checking and adjusting cam timing, check piston-to-intake- and –exhaust-valve clearances. The piston is closest to the valve immediately before, during and immediately after TDC valve overlap. As a rule, the minimum allowable piston-to-valve clearance is 0.090—0.100 in. at the exhaust valve and 0.070—0.080 in. at the intake valve.

Additional clearance is needed at the exhaust valve because it is closing as the piston comes up and "chases" it. Inertia during high-rpm closing unloads the valve train to varying degrees, thus reducing exhaust-valve-to-piston clearance.

Check exhaust-valve clearance from 20° BTDC to TDC in 5° increments. Then, check intake-valve clearance from TDC to 20° ATDC, also in 5° increments. Some cam grinders provide the crank positions at which the valves are closest to the pistons and the points at which to check these clearances.

One way to check clearance is to measure installed valve-spring height, then compress the rocker-arm tip and valve assembly until the valve head touches the piston top and remeasure valve-spring height. The difference between the two dimensions is piston-to-valve clearance. Another way to measure this clearance is directly through the spark-plug hole, either with special feeler gages or with a malleable substance such as piece of 0.120-in.-diameter solder positioned between the piston top and valve head. When using the solder, rotate the crankshaft through the crucial zone, then remove and mike the solder. Its squeezed thickness gives a direct piston-to-valve clearance reading.

Record minimum clearance so you can consider necessary modifications, if any are needed. If there's not adequate piston-to-valve clearance, you'll have to lower the valve pockets in the piston tops by flycutting.

Fit Intake/Exhaust-Manifold Gasket—Most racing or high-performance engines use compressed fiber, reusable-type manifold gaskets. This type of gasket can be easily trimmed to match port shapes. Because this trimming process is messy, it shouldn't be done on a final-assembled "clean" engine. So, now's the time to do it.

While the cylinder head is on the block, install the intake/exhaust-manifold gasket. Align the gasket with the ports and secure it in position. Trim away excess that overhangs the port openings with an X-acto knife or equivalent.

Conclusion—Providing you performed all procedures outlined in this chapter, you can disassemble the engine. To prepare the engine for final assembly, clean each of the components and reorganize them. Slip a plastic bag back over the block and cover the other components to keep them dust-free while they await assembly.

CHAPTER EIGHT
Engine Assembly

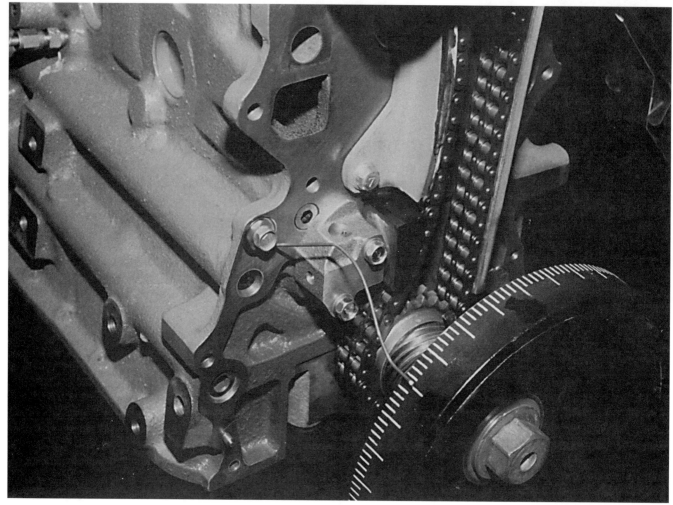

When engine goes together at this stage it's for real. It'll not come apart again unless there's an engine failure or it's time for inspecting and rebuilding. If you haven't checked clearances and cam timing, do it before proceeding.

If you haven't checked or adjusted all critical and non-critical clearances, and inspected all components, do so *before* you begin final engine assembly. I have written this chapter with the assumption that you've already done so. All components must be *thoroughly cleaned*.

Organize your work area. The workbench should be clean and covered with newspaper, and you should have a fresh supply of clean solvent—preferably in a cleaning tank—and a roll of lint-free paper towels nearby. And, don't forget to drain all water from your air-compressor tank. Assuming you've met all the above conditions, remove the plastic bag from your clean bare block and begin assembling your engine.

Following is a step-by-step procedure for assembling your engine. Do each step as I recommend and chances are you won't go wrong. Good luck!

BLOCK

• Install the oil-gallery plugs and fittings. Epoxy all fittings and plugs on the outside of the engine where seepage may cause a problem. Any external oil leakage on a race engine is unacceptable. It may affect clutch operation or leak on the exhaust system. Minor oil seepage inside the engine flows harmlessly back to the oil sump.

Caution: Be sure the plug in the front of the main oil gallery installs flush or below the front face of the block.

• Install the water-jacket core plugs with epoxy. Drive each plug into place with a pilot/driver that fits the plug without distorting it. Steel core plugs are OK for race

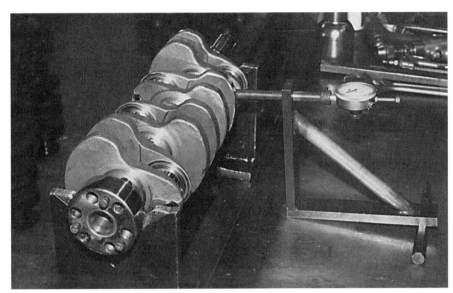
Straightness of race-prepped L20B crankshaft is checked one last time before it's installed. Pilot bushing will be installed next.

Among materials required for engine assembly, you'll need something to lube the cam. MoS_2 is a good choice for this.

Reciprocating components of L28 GTU engine have been fitted, numbered and cleaned prior to final assembly.

engines because corrosion should not be a problem. Rotate the block upside down.
• On wet-sump engines, install the crankcase-vent screen and baffle in the block.
• Install the crankshaft oil-passage plugs, three keys into crank-snout keyways, and the pilot bushing or needle bearing at the flywheel end of the crank.
• With a clean, lint-free paper towel, wipe any oil from the main-bearing bores and mating surfaces of the block and caps.
• Oil the bolt threads and underside of the heads.
• Wipe off the back of the bearings and, using the numbers you put on the backside of the main-bearing shells as reference, *insert* the bearings into the block and the numbered caps.
• Apply a moderate amount of assembly lube to the bearings. Don't forget to lube the thrust faces of the center main. Spread the lube over bearings with your fingers.
Note: A 50/50 mixture of a cam-manufacturer's assembly lube—usually molybdenum disulphide—and motor oil in an oil squirt can works well for prelubing bottom-end components. Also, products such as Lubriplate® make excellent assembly lubes.
• Carefully lower the crankshaft into the block and onto the bearings.
• Oil or grease the rear-main-seal lip and install it. The seal lip should point toward the crankshaft and the rear face of the seal should be flush with the rear face of the cylinder block.
• Install the main-bearing caps on their journals. The arrows point forward and

ENGINE-ASSEMBLY TORQUES
ft-lb (kg-m)

Camshaft-sprocket bolt	100-108 (13.8-14.9)
Camshaft thrust-plate bolts	4-7 (0.5-1.0)
Camshaft-tower bolts	10-12 (1.3-1.6)
Chain-guide bolts	7(1)
Chain-tensioner bolts	7(1)
Connecting-rod bolts 8mm/9mm	20-24 (2.8-3.3)/33-40 (4.4-5.4)
Crank-damper/pulley bolt	100-108 (13.8-14.9)
Cylinder-head bolts	
L13, 16, 20A	40 (5.5)
L18, 20B, 24, 26, 28	61 (8.5)
L28 Turbo	65 (9.0)
Flywheel bolts	100-108 (13.8-14.9)
Front-cover bolts 6mm/8mm	4-6 (0.6-0.8)
8mm	9-11 (1.2-1.5)
Intake/exhaust-manifold	
bolts: 8mm/10mm	9-12 (1.2-1.6)/25-35 (3.4-4.7)
nuts	9-12 (1.2-1.6)
Main-bearing-cap bolts	30-40 (4.5-5.5)
Oil-pan bolts	5-7 (0.7-1.0)
Oil-pump bolts	9-11 (1.2-1.5)
Oil-pump-pickup bolts	8-10 (1.1-1.4)
Rocker-arm-cover bolts	5-6 (0.7-0.9)
Rocker-arm-pivot locknut	40-44 (5.5-6.0)
Rocker arm pivot sleeves	100 (13.8)
Spark plugs	12-15 (1.2-2.0)
Spray bar bolts	5-6 (0.7-0.9)

Never guess when final-tightening bolts or nuts. Torque them to correct specifications and, if required, in specified sequence. If specifications are not given in chart at top, torque fasteners according to specifications below.

STANDARD TORQUE-NMC BOLTS

Bolt Dia. (mm)	Grade*	Thread Pitch (mm)	Tightening Torque (ft-lb)	(kg-m)	(N-m)
6	4	1.0	2-3	0.3-0.4	3-4
8	4	1.0-1.25	6-8	0.8-1.1	8-11
10	4	1.25-1.5	12-16	1.6-2.2	16-22
12	4	1.25	22-30	3.1-4.1	30-40
12	4	1.75	20-27	2.7-3.7	26-36
14	4	1.5	34-46	4.7-6.3	46-62
6	7	1.0	4-5	0.6-0.7	6-7
8	7	1.0-1.25	10-13	1.4-1.8	14-18
10	7	1.25-1.5	19-27	2.6-3.7	25-36
12	7	1.25	37-50	5.1-6.9	50-68
12	7	1.75	33-45	4.6-6.2	45-61
14	7	1.50	56-76	7.7-10.5	76-103
6	9	1.0	5.8-8.0	0.8-1.1	8-11
8	9	1.0-1.25	14-19	1.9-2.6	19-26
10	9	1.25-1.5	27-38	3.7-5.2	36-51
12	9	1.25	53-72	7.3-9.9	72-97
12	9	1.75	48-65	6.6-9.0	65-88
14	9	1.5	80-108	11.1-15.0	109-147

*Bolt grade embossed on bolt head.

the numbers correspond to crankshaft bearing-journal numbers.

Caution: Take care while inserting the rear cap in its register. The bearing shell may drop out of position. Also, align the rear surfaces of the seal and cap so they'll be flush with the rear face of the block after the cap bolts are tightened.

- Align the center-bearing thrust faces by forcing the crank back and forth in the block. Do this with a pry bar or large screwdriver between a crank throw and main-bearing web. Hold the crank forward or rearward in the block and simultaneously tighten the center main-cap bolts.
- Torque the main-bearing-cap bolts to specification in three equal steps. Torque standard bolts 33—40 ft-lb (4.5—5.5 kg-m); LD28 diesel bolt torque spec is 51-61 ft-lb (7.0—8.4 kg-m). All bolts should feel the same when torqued. Any bolt that requires more rotation or feels "softer" than the others is probably stretching and should be discarded. Check crankshaft rotation after the main-cap bolts are final-torqued. It should spin freely.
- Assemble each connecting rod, piston and pin using the reference numbers that indicate which cylinder each goes to. To assemble a rod and piston, oil the pin and pin bore. Put the rod oil hole, if so equipped, on the dome or spark-plug side of the piston. If retaining rings are used to retain the pin, install them. Remember, circlips are installed with the rounded edge against the pin; sharp edge installs away from pin.
- Install the piston rings in the correct order and, if required by design, with the top sides up. Follow the ring-manufacturer's instructions. It's best to use a ring expander to install the rings.
- If you're using standard Nissan rod bolts, install new ones.
- Insert the connecting-rod bearings—as numbered—into their respective rods and caps. Apply a moderate amount of bottom-end lube to the bearings. Install sleeves over the rod bolts. Two 2- or 3-in. lengths of fuel hose will do.
- Rotate the crankshaft so journal number-1 is down.
- Apply a small amount of oil to the piston skirts.

Engine Assembly 87

Align thrust-bearing-insert halves by levering crank back and forth in block before tightening center main-cap bolts Tighten these bolts while forcing crank forward. Final-torque main-cap bolts 30—40 ft-lb (4.5—4.4kg-m), If you're using LD28 main cap bolts, final torque to 51—61 ft-lb (7.0—8.4kg-m). Tom Monroe photo.

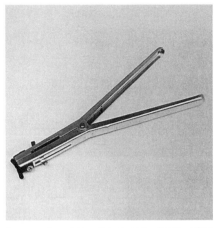

Inexpensive piston-ring spreader beats using thumbs for installing piston rings. Tom Monroe photo.

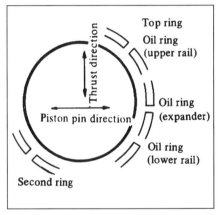

Before compressing rings, position them as shown. Oil-ring expander/spacer goes at back side of piston—opposite piston notch or arrow. Position top oil-ring rail under top compression ring and bottom rail in the same position on the opposite side of the expander. Drawing Courtesy Nissan.

Opinions vary as to exactly where the end gaps of the first and second compression rings should be positioned. But, there is agreement on one point: End gaps of the two compression rings should not be installed in line with the thrust face of the piston and not be less than 120° apart from each other.

• Compress the rings with a ring compressor, then install each piston assembly into its bore. Check that the piston dome is toward the spark-plug side of the bore—side opposite the distributor—and that the front of the piston is forward. With the block rotated upright, guide the piston-and-rod assembly into its bore.

If the piston binds or hangs up as you push it into its bore, STOP! A ring is probably hanging up at the top edge of the bore. Remove the piston-and-rod assembly and recompress the rings. Forcing the piston can easily break a ring or damage a piston.

Once the piston is into the bore, set the ring compressor aside and turn your attention to aligning the rod big end with the rod journal. Push the piston down the bore and guide the rod onto its journal. If you didn't put sleeves on the rod bolts, don't let the rod or its bolts scratch the journal.

• After oiling the insert, install the cap on its rod. The cap and rod numbers or bearing tangs should align with each other. Oil the rod-nut or -bolt faces and threads. Install them and evenly tighten until snug. Torque the nuts or bolts to specification: 33—40 ft-lb (4.5—5.4 kg-m) for standard 9mm rod bolts; 20—24 ft-lb (2.8—3.3 kg-m) for 8mm rod bolts. For special bolts, use the manufacturer's torque spec.

Note: Now that you've installed a rod and piston, the crankshaft will be slightly more difficult to rotate. To make it easier, install a stock damper/pulley bolt with its washer and tighten against the crank snout. You can turn the crank with a wrench on the bolt.

• Rotate the crankshaft and check for

Assembled short block is ready for cylinder head to be installed. Copper O-rings surrounding bores improve head-gasket seal. Note cleanliness of block, bores and pistons. Keep it clean!

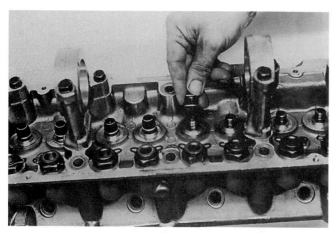
Begin valve installation by installing seals. Tom Monroe photo.

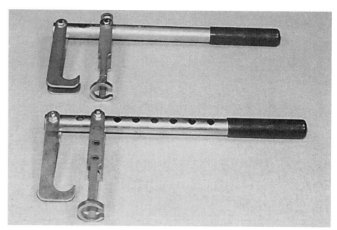
Either valve-spring compressor tool can be used to compress valve springs for installing rocker arms. Bottom tool is for all Nissan/Datsun OHC engines.

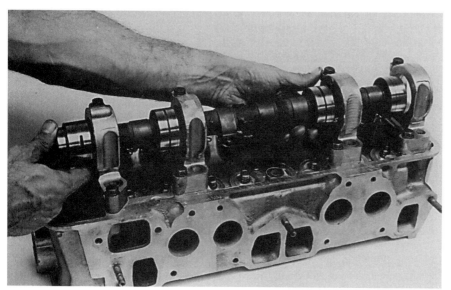
Ultimate test of cam-tower alignment is that cam rotates freely once it is installed. Tom Monroe photo.

binding. Position the number-2 rod journal down and continue installing the piston-and-rod assemblies.

CYLINDER HEAD
- Install all valve-guide seals.
- Assemble the valves and components in *their* positions: the number-1 valve with springs, shim(s), retainer and keepers in number-1 guide.
- Oil the number-1 exhaust-valve stem and insert it into its guide. Install the valve-spring shim(s), the springs on the shim(s), then the spring retainer on the springs. Using a spring compressor, compress the spring/retainer assembly. Use caution not to damage any part of the head with the spring compressor. Install the two keepers on the valve stem. A little grease on each will help hold them in place.

As you cautiously loosen the compressor, check that the keepers stay in place as the retainer seats on them. To check seating of the retainer keepers, moderately strike the top of the retainer with a hammer. If a keeper is not seated, it may pop loose. Also, make sure the spring shims stayed in position during spring installation. Double-check all valve and spring parts, then proceed with installing the number-1 intake valve.
- Repeat the installation process on the remaining valves.
- If you haven't installed the rocker-arm-pivot assemblies, do so now. Torque all pivot sleeves 100 ft-lb (13.8 kg-m). The factory drawings apply different names to the pivot sleeves. They are called *Bushing, rocker pivot* in the page 89 drawing, item 19. The same parts are called *Rocker-pivot mount* on the page 90 drawing, item 15.
- Install a rocker-pivot spring retainer on each pivot sleeve. It should be snug on the sleeve.
- Using the numbers on the cam towers for reference—you should've numbered them during cylinder-head disassembly—install the cam towers and, if required, cam-tower shims. There should be two hollow dowels for each cam tower. They install in counterbores in two bolt holes for each of the cam towers. Install any missing dowels *before* the cam towers.

Oil the threads and underside of each cam-tower-bolt head. Thread-in the bolts and loosely tighten them. After oiling the journals, install a camshaft that you know is straight. It doesn't have to be the one you're going to use. The cam should rotate freely. Gradually tighten all cam-tower bolts and recheck cam rotation. Final-

Engine Assembly

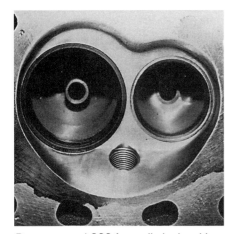

Race-prepared SSS four-cylinder head has been thoroughly cleaned prior to assembly. If you glass-beaded cylinder head, it must be washed numerous times in soapy water to remove all traces of glass particles. Pay particular attention to valve guides.

torque cam-tower bolts 10—12 ft-lb (1.3—1.6 kg-m) and recheck rotation.

If the cam binds while torquing the cam-tower bolts, slightly loosen the bolts at one cam tower and check rotation. Retighten those bolts and go to the next cam tower until the suspect tower is found. If while loosening the bolts, rotating the cam and retorquing the bolts doesn't reduce or eliminate the bind, slightly loosen the bolts again and moderately strike the tower vertically with a plastic mallet. Retorque the bolts and recheck cam rotation. Continue the process until the cam rotates freely. Remove the cam.

INSTALL CYLINDER HEAD

- If your block has been modified for O-ring wire, install the wire. Use a plastic mallet to tap the wire into the groove. Position the wire ends away from water passages and adjacent cylinder bores. Be sure the ends butt. There should be no gap or overlapping at the ends.
- Rotate the crankshaft to bring cylinder-1 piston to TDC of its compression stroke.
- Clean the block and cylinder-head gasket surfaces with an evaporative, non-petroleum-based solvent such as lacquer thinner.

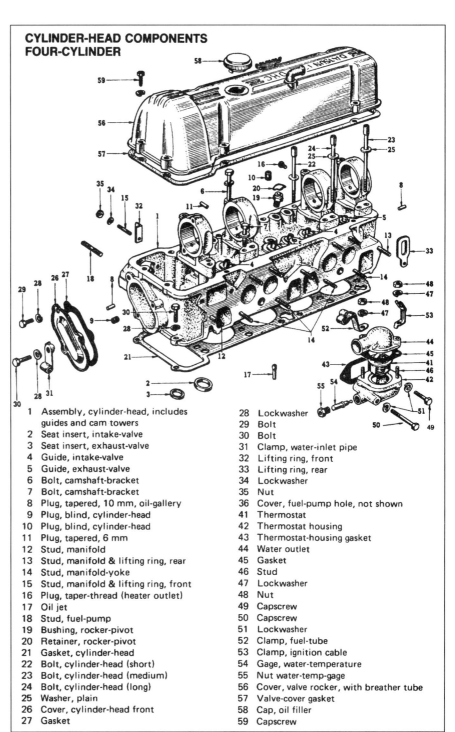

CYLINDER-HEAD COMPONENTS FOUR-CYLINDER

1. Assembly, cylinder-head, includes guides and cam towers
2. Seat insert, intake-valve
3. Seat insert, exhaust-valve
4. Guide, intake-valve
5. Guide, exhaust-valve
6. Bolt, camshaft-bracket
7. Bolt, camshaft-bracket
8. Plug, tapered, 10 mm, oil-gallery
9. Plug, blind, cylinder-head
10. Plug, blind, cylinder-head
11. Plug, tapered, 6 mm
12. Stud, manifold
13. Stud, manifold & lifting ring, rear
14. Stud, manifold-yoke
15. Stud, manifold & lifting ring, front
16. Plug, taper-thread (heater outlet)
17. Oil jet
18. Stud, fuel-pump
19. Bushing, rocker-pivot
20. Retainer, rocker-pivot
21. Gasket, cylinder-head
22. Bolt, cylinder-head (short)
23. Bolt, cylinder-head (medium)
24. Bolt, cylinder-head (long)
25. Washer, plain
26. Cover, cylinder-head front
27. Gasket
28. Lockwasher
29. Bolt
30. Bolt
31. Clamp, water-inlet pipe
32. Lifting ring, front
33. Lifting ring, rear
34. Lockwasher
35. Nut
36. Cover, fuel-pump hole, not shown
41. Thermostat
42. Thermostat housing
43. Thermostat-housing gasket
44. Water outlet
45. Gasket
46. Stud
47. Lockwasher
48. Nut
49. Capscrew
50. Capscrew
51. Lockwasher
52. Clamp, fuel-tube
53. Clamp, ignition cable
54. Gage, water-temperature
55. Nut water-temp-gage
56. Cover, valve rocker, with breather tube
57. Valve-cover gasket
58. Cap, oil filler
59. Capscrew

Typical L-series four-cylinder head and related components less camshaft and valve train. Drawing courtesy Nissan.

- Double-check that the two cylinder-head dowels are in place. If not, install them now. Each goes in a countersunk hole at a cylinder-head bolt hole.

- Install the head gasket onto the cylinder block, aligning it on the dowels. Standard Nissan gaskets should be installed dry, without adhesives. If the engine uses

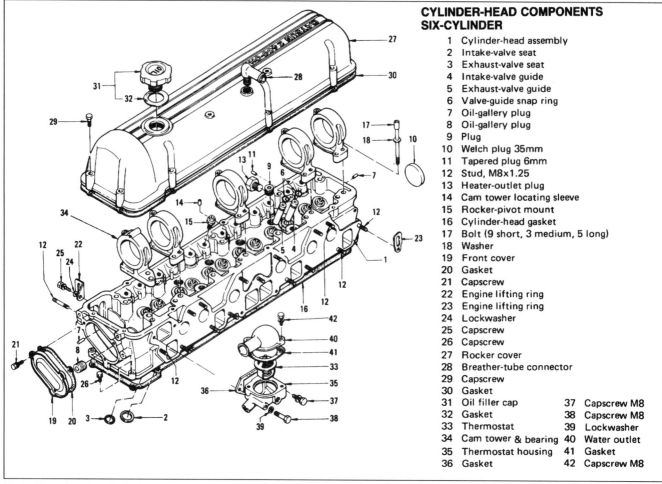

CYLINDER-HEAD COMPONENTS SIX-CYLINDER

1. Cylinder-head assembly
2. Intake-valve seat
3. Exhaust-valve seat
4. Intake-valve guide
5. Exhaust-valve guide
6. Valve-guide snap ring
7. Oil-gallery plug
8. Oil-gallery plug
9. Plug
10. Welch plug 35mm
11. Tapered plug 6mm
12. Stud, M8x1.25
13. Heater-outlet plug
14. Cam tower locating sleeve
15. Rocker-pivot mount
16. Cylinder-head gasket
17. Bolt (9 short, 3 medium, 5 long)
18. Washer
19. Front cover
20. Gasket
21. Capscrew
22. Engine lifting ring
23. Engine lifting ring
24. Lockwasher
25. Capscrew
26. Capscrew
27. Rocker cover
28. Breather-tube connector
29. Capscrew
30. Gasket
31. Oil filler cap
32. Gasket
33. Thermostat
34. Cam tower & bearing
35. Thermostat housing
36. Gasket
37. Capscrew M8
38. Capscrew M8
39. Lockwasher
40. Water outlet
41. Gasket
42. Capscrew M8

Typical L-series six-cylinder head and related components less camshaft and valve train. Drawing courtesy Nissan.

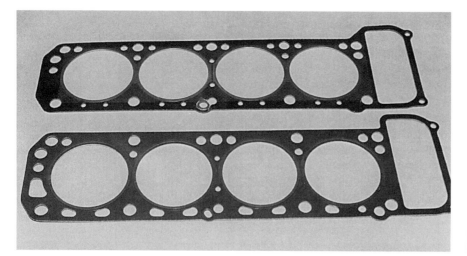

Top head gasket is standard Nissan/Datsun part that should be installed dry. Bottom gasket is 11044-U0820 competition gasket that requires a sealer such as Hi-Tack or similar.

Note: An upgraded head bolt was supplied on the 280ZX Turbo. It is stronger and will fit all L-series engines. It can be identified by a ring recess on top of the bolt outside of the hex socket. Part No. is 11056-P7600 (short bolt) and 11059-P7600 (long bolt).

17mm crowfoot wrench at left can be used to loosen or torque rocker-arm-pivot locknuts. The 1/2-in.-drive, long-shank 10mm Allen hex at right is for installing and removing socket-head bolts such as the head bolts. Both are available from Nissan Motorsports.

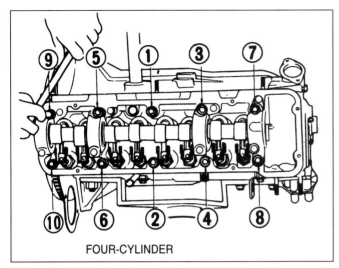
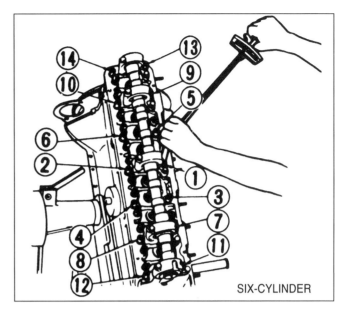

Torque head bolts in order and in 10-ft-lb (1.3-kg-m) increments. Final-torque L28 Turbo bolts 65 ft-lb (9.0 kg-m). Drawings courtesy Nissan.

a sealing-ring-type head gasket, position the sealing rings in the cut grooves, and apply a suitable adhesive, such as Hi-Tack®, to the gasket. If an aftermarket gasket is being used, be sure to follow the gasket manufacturer's recommendations. **Note:** If you're using sealing rings, turn to page 17 for how to install the rings and gasket.

- Install the head onto the block and gasket. Register the head on the alignment dowels.
- Clean the head-bolt washers in lacquer thinner or equivalent and clean the head-bolt washer surface on the cylinder head.
- Oil the head-bolt threads and underside of the bolt heads. Anti-seize compound is also good for lubricating the underside of the bolt heads.
- Place the head-bolt washers concave side down on the head and install the head bolts. Starting at the two center bolts, work toward the ends of the head, alternately snugging the bolts. Use a long-shank 10mm Allen socket and, to make the job go faster, a speed handle.
- Torque the head bolts in order and in 10 ft-lb (1.3 kg-m) increments. Final-torque them 65 ft-lb (9.0 kg-m). Use an accurate torque wrench, preferably a *wand type,* because they are accurate. Head-bolt torque is critical, especially with O-ring or *sealing-ring-type* head gaskets, because the O-ring wires or sealing rings are compressed along with the head gasket.

INSTALL VALVE TRAIN & CAM DRIVE

- Coat the cam-bearing journals with assembly lube and install the cam. Check the cam for easy rotation.
- Install the cam retainer/thrust plate. The cam-timing dash mark installs to the front. Apply blue Loctite to the bolt threads. Torque the thrust-plate bolts 5—6 ft-lb (0.7—0.8 kg-m).
- Position and install the tension side—straight—chain guide. Use blue Loctite on the bolts and torque them 7 ft-lb (1.0 kg-m).
- Install the slack-side chain guide, but leave the bolts loose.
- Rotate the cam so the sprocket dowel is at the top, or 12 o'clock, position.
- Install the crankshaft sprocket so the chain aligning dot is at the front. Cylinder-1 piston should be at TDC.
- With the timing chain draped over it, install the cam sprocket and chain. Align one bright link to the crankshaft-sprocket dot. It will be at about the 4 o'clock position when installed. Align the second bright link to the cam-sprocket dot, as determined during fit-up/mock-up, page 81. The link/dot position on the cam

Cam-drive components in as-installed positions. Bright link at bottom aligns with crank-sprocket timing mark. Bright link at top aligns with cam-sprocket mark you determined to be correct when timing cam. Tom Monroe photo.

Dial-In-Cams adjustable cam sprocket is popular on L-series race engines because cam timing is easily adjusted. If you use this setup, keep bolts tight.

With bright links (arrows) aligned with sprocket timing marks, slide crank sprocket on and start cam-sprocket bolt into nose of cam. Cam may need to be rotated slightly to align dowel with hole in sprocket. Tom Monroe photo.

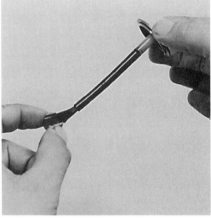

To double-check lobe wipe pattern, coat rocker-arm contact pads with layout blue as is being done with lash cap. Tom Monroe photo.

sprocket will be at about 2-o'clock when the corresponding dowel hole is fitted to the cam dowel.

- Install the cam sprocket to the cam nose. Apply Loctite to the cam-sprocket bolt, thread it in the nose of the cam and torque 100—108 ft-lb (13.8—14.9 kg-m). On a four-cylinder engine, hold the cam when torquing the sprocket bolt with an adjustable wrench on the cast-in square bosses near the center of the cam.
- Install the chain tensioner with Loctite on the bolt threads. Before tightening, align the tensioner and the curved slack-side chain guide simultaneously. Correct tensioner and guide positioning will remove slack from the chain and align the chain so it passes over the center of the tensioner shoe and onto the guide in a smooth arc. If chain contact on the tensioner shoe is not centered, it will damage or wear the tensioner-shoe face. Torque the tensioner and curved-guide bolts 7 ft-lb (1.0 kg-m).
- Oil each valve-stem tip and place the selected lash pads into the corresponding valve-spring retainers.
- Coat the rocker-arm contact pads with machinist bluing or a permanent felt-tip maker.
- Rotate the crankshaft and camshaft so the number-1 cam lobe points up, away from the head. Starting with valve-spring number-1, compress the valve spring with a compressor and install its rocker arm. Adjust cold valve-lash clearance according to the cam-manufacturer's specifications. Torque the rocker-arm-pivot locknut 40—44 ft-lb (5.5—6.0 kg-m). Recheck lash after securing the locknut.
- Rotate the camshaft 360°, or the crank 720°, and return the number-1 cam lobe to the up position. Without removing the rocker arm, check the cam-lobe wipe pattern. If it's OK—centered or in the position you want—coat the number-1 lobe with assembly lube and continue installing rocker arms and checking the wipe patterns.
- Install the rocker-arm-lash, "mousetrap"—springs.
- On six-cylinder engines, install the spray bar. Torque the retaining bolts 5—6 ft-lb (0.7—0.8 kg-m). Use caution when installing and torquing the bolts to ensure a good seal at the spray bar-to-cam tower gaskets.

Engine Assembly

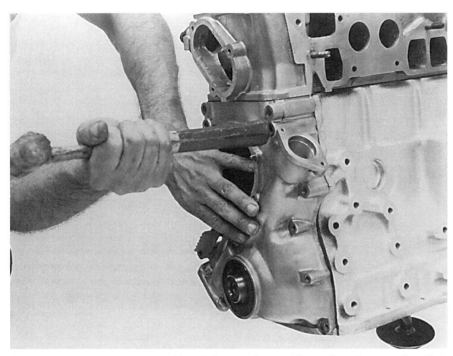

Front cover can go on once cam-drive chain, sprockets, guides and tensioner are installed. Cover is tapped in place over tapered alignment dowels. Tom Monroe photo.

Distributor/oil-pump drive spindle is indexed in correct position. Drive-tang offset is to front and in 11:25 position, or directed tangent to backside of top distributor-adapter mounting bolt.

INSTALL EXTERNAL PARTS

- Install a new seal in the front cover. The front face of the seal should be installed flush with the front edge of the seal bore and the seal lip should point toward the engine. Oil or grease the seal lip.
- Apply anti-seize compound to the front-cover gaskets and to the front-cover portion of the head gasket.
- Slide the bronze oil-pump/distributor drive gear onto the crankshaft snout. The end with the large chamfer goes on first. **Note:** Do not install a crankshaft oil slinger.
- Check that the forward-most crank key is installed and that the two front-cover dowels are in the front face of the block. One dowel installs at the upper left and the other at the lower right.
- Install the front-cover gaskets and front cover. Check that the side gaskets stayed in place and that the front edge of the head gasket is in position. Loosely install the two head-to-front cover bolts that thread down through the flange at the front of the head into the front cover. Install and torque the remaining bolts. Torque 8mm bolts 9—11 ft-lb (1. 2—1.5 kg-m) and 6mm bolts 4—6 ft-lb (0.6—0.8 kg-m).
- Apply *green* Loctite to the crank snout and the damper/pulley bore. Install the damper/pulley. Use caution to align the damper/pulley with the crank key. Apply *red* Loctite to the crank-damper/pulley bolt threads. If you're using a stock damper/pulley bolt, torque it 100—108 ft-lb (13.8—14.9 kg-m). But, if you are installing a non-stock damper/pulley, torque its bolt to the manufacturer's recommendations.
- Double-check alignment/timing of the crank-damper/pulley timing mark to the timing indicator and recheck cam timing. If cam timing requires adjustment, use a *chain wedge* to lock the chain in place. Drive the wedge in place between the tight and slack sides of the chain, then remove the cam sprocket.

Advance or retard the cam as required by aligning the bright chain link with the desired sprocket timing mark. Hold the sprocket against the cam nose and rotate the cam to align the dowel with the dowel hole. Reinstall the cam sprocket and sprocket bolt and remove the chain wedge.
- Install the cylinder-head inspection plate.
- Rotate the crankshaft to bring cylinder-1 piston up to TDC on the compression stroke. Apply anti-seize compound to both sides of the oil-pump gasket—block-off plate if you're using a dry-sump system—and fit it to the pump. Slide the oil-pump/distributor-drive spindle in the oil pump or plate. Index the distributor-drive tang at the upper end of the spindle.

When correctly indexed and installed, the spindle will be offset forward and in the 11:25-o'clock position. Also, a line projected from the tang will be tangent to the backside of the top distributor-mounting bolt hole. There is a nice drawing of this in any L-series service manual. It helps to see this. If the engine will have a wet sump, fill the oil pump with high-viscosity oil—about 50W—and install the pump and spindle. If the engine is equipped with a dry-sump pump, install the block-off plate in the original oil-pump location with the spindle, page 100. Torque the oil-pump or bolts 9—11 ft-lb (1.2—1.5 kg-m).

Dry-sump oil pump is driven by Gilmer type cog belt. Correct tension is achieved when belt can be twisted a full 90°.

Black felt-tip-pen marks on each spark-plug body indicates ground-electrode position. Ground electrodes are indexed toward intake-valve side of combustion chamber. Electrode is also clear of "pop-up" piston dome.

- Glue the compressed-fiber, reusable pan gasket to the oil pan with weatherstrip adhesive. Coat the block side of the gasket with anti-seize compound.
- Invert the engine assembly. If the engine uses wet-sump lubrication, install the oil-pump pickup with a new gasket. Torque the pickup bolts 8—10 ft-lb (1.1—1.4 kg-m).
- Install the oil pan. First install all bolts loose, then tighten them in a crisscross pattern. Final-torque oil-pan bolts 5—7 ft-lb (0.7—1.0 kg-m). Bolt tension will drop as the pan gasket relaxes, so retorque the bolts several times.
- On wet-sump engines, install the drain plug and torque it 25—30 ft-lb (3.5—4.1 kg-m).
- Rotate the crankshaft to check for any interference between the crankshaft or connecting rods and oil pan.
- Rotate the engine so the cylinder head is up.
- Glue the valve-cover gasket to the valve cover. Oil the cylinder-head side of the gasket. Install the valve cover and valve-cover bolts finger-tight, then final-torque them 5—6 ft-lb (0.7—0.8 kg-m).
- Rotate the crankshaft assembly to 35° BTC of the compression stroke for cylinder 1. To check the position of the cylinder-1 intake and exhaust cam lobes to confirm cylinder 1 is on its compression stroke, remove the valve-cover oil-filler cap. You'll be able to see the cam lobes. Both should be up.

Install the distributor with the rotor pointing to the number-1 firing position. Use the distributor cap as reference to mark the position of the cylinder-1 spark-plug-wire terminal on the distributor housing.
- While the crankshaft is still at 35° BTC on cylinder-1 compression stroke, position the crank-mounted magnetic-pickup assembly if so equipped. The trailing edge of the paddle should align with the center of the magnetic-pickup assembly. Also, there should be 0.035—0.050-in. (0.89—1.27mm) clearance between the pickup and all of the paddles.
- Before you install the spark plugs, check their gap against the manufacturer's specifications. Most fine-wire racing plugs are gapped to 0.021 in. (0.53mm). Rotate the crankshaft so none of the pistons are at TDC. This is because some dome-top pistons require that the spark plugs be indexed—installed so the ground electrode is away from the dome. Otherwise, the piston dome will contact the spark-plug ground electrodes.

If the spark-plugs must be indexed, remove the standard sealing washers and install copper spacers in their place.

Using a permanent-ink felt-tip marker, draw a line from the ground electrode up the body of the plug. Install the copper spacer and apply anti-seize compound or oil to the plug threads. If you're using anti-seize compound, don't get it on the first two threads. If the compound gets on the plug electrode, it will short the plug.

Install the spark plugs and tighten gently. Torque each plug 12—15 ft-lb (1.6—2.0 kg-m). Check the position of the index mark on each spark-plug. It should be in the 10—11- or 1—2 o'clock position. If it's not, remove the plug and install it in another hole. Continue this process until all plugs are positioned correctly and torqued to spec.
- Install the thermostat-housing assembly. Apply anti-seize compound to the gaskets. Use caution when installing the long front horizontal bolt into the cylinder head. If this bolt is too long, it will contact the backside of the tight-side chain guide and bend it.

To make sure this doesn't happen, thread in the bolt before you install the

Engine Assembly

Small hole (arrow) in manifold-gasket surface centered below intake ports is for positioning dowel. Companion hole is in manifold and gasket. Dowel aligns intake ports with manifold and gasket.

Don Devendorf's Turbo-ZX GTO race car uses triple-disc Quarter Master clutch assembly. Clutch discs are centered to crankshaft with transmission input shaft. Refer to clutch chart, in glossary for specific applications.

thermostat housing. Run the bolt in until it stops, then measure between the head and the underside of the bolt head. If this distance is less than the thickness of the thermostat-housing boss, it's OK. Install the modified thermostat or flow restrictor and the anti-seize-coated gasket.

The restrictor can be either a plate with a 5/8-in. hole drilled in it, installed at the standard thermostat position, or as a unit in the water-outlet neck. A standard thermostat can be made into a flow restrictor by removing its spring.

- Position the compressed-fiber-type intake/exhaust-manifold gasket to the side of the cylinder head. If the gasket has not yet been trimmed, cut it as necessary to match the shapes of the ports with a sharp, fine-point knife. Remove the gasket and apply anti-seize compound to both sides. Reinstall the gasket, then the exhaust header.

Align the gasket with the intake ports and install the front, center and rear nuts on the studs that secure just the header. Install the engine-lift bracket—slinger—on the rear-most stud. This should hold the gasket in position. Install the intake manifold. Install the manifold *yoke washers* convex side against flanges and nuts. Tighten all nuts and bolts evenly 12—15 ft-lb (1.6-2.0 kg-m).

- Remove the engine assembly from the engine stand and support it firmly on the floor or bench. You can't install the flywheel or clutch with it mounted on the stand.
- Before installing the flywheel, check the pilot bushing/bearing in the end of the crankshaft. Apply a small amount of grease to the bushing/bearing. Wipe clean the crankshaft flange and mating surface of the flywheel. Install the engine plate.

If the crankshaft and flywheel were balanced as an assembly, align the index mark on the flywheel with the one on the crankshaft flange and install the flywheel. Always use new bolts with Loctite on the threads to secure the flywheel. Tighten the bolts gradually in a crisscross pattern and final-torque them 100—108 ft-lb (13.8—14.9 kg-m). Strike the flywheel-bolt heads with a hammer and retorque them to the same spec.

- Before installing the clutch assembly, clean the flywheel, disc(s) and pressure-plate friction surfaces with an evaporative, alcohol-base cleaner. Align all balance index marks during installation. Use an aligning tool or transmission input shaft to align the clutch disc(s).

Gradually tighten the pressure-plate bolts in a crisscross pattern, then remove the aligning tool or input shaft. Torque the securing bolts to 11—16 ft-lb (1.5—2.2 kg-m), if stock bolts are used. Reinsert the aligning tool to check disc alignment. If the tool doesn't fit, loosen the pressure-plate bolts, realign the disc(s) and retighten the bolts. Recheck disc alignment.

CHAPTER NINE
Lubrication

Jim Fitzgerald in his immaculately prepared SCCA GT-2 280ZX: Dry-sump lubrication system ensures continuous oil pressure to engine during high-g braking and cornering.

Virtually any L-series oil pump will install on any L-series engine. Also, nearly all L-series oil pumps have the same part number, 15010-21001. There are two exceptions. The first exception is the 280ZX-T (T for turbocharged) oil pump 15010-S8000. This was the first and only change to an original-equipment L-series oil pump since the introduction of the L-series engine. The second exception is Nissan Motorsports' high-pressure oil pump 15010-A1110.

The output of standard and competition pumps is 0.49 fluid ounces (14.5 ml) per pump revolution. The output of the 280ZX-T oil pump is 0.55 fl-oz (16.4 ml) per pump revolution. The higher volume is for additional oil to the turbocharger. This volume-flow increase was accomplished with 5mm (0.20-in.) longer gears, or 35mm (1.378 in.) versus 40mm (1.575 in.). Externally, the *turbo* oil pump is unchanged.

OIL-PUMP PRESSURE ADJUSTMENT

Except for the pressure-relief-valve springs, the standard pump and Motorsports oil pump are virtually identical. The Motorsports pump high-pressure-relief springs—outer spring 15133-22010 and inner spring 15133-E4620—are available separately. Installing them will convert a standard pump to what amounts to a Motorsports oil pump. High-pressure relief-valve springs also fit the 15010-58000 high-volume turbo pump to give both high volume and high pressure for the most demanding applications.

Because the L-series engine has an externally mounted oil pump—it's at the lower right corner of the front cover—access to it is easy. Oil pressure is adjusted by mixing and matching relief-valve springs and installing shims or small washers under the springs.

Gain access to the relief valve and springs by removing the large 22mm hex

22mm pressure-relief-valve plug in stock oil-pump cover can be removed with pump mounted on engine. This allows relatively easy changes to oil pressure.

Nissan Motorsports' oil-filter adapter allows easy plumbing of oil cooler with wet-sump lubrication system.

plug. The plug installs in the bottom of the pump from the rear.

Oil Pressure—There's no easy answer as to how much oil pressure to run. It's largely a matter of personal preference. For stock engines, Nissan service calls for 11-psi oil pressure at idle. The stock relief-valve opening pressure varies between 50 and 71 psi, or 3.5—5.0 kg/cm².

On wet-sump oil systems for performance-rally engines, Nissan Japan recommends 85±7-psi (6±0.5-kg/cm²) maximum oil pressure. On most engines modified for the street, where maximum engine speed seldom exceeds 7000 rpm, 60—70-psi (4.25—5-kg/cm²) maximum oil pressure is desirable. On most engines that will exceed 7000 rpm frequently, 70—100-psi (5—7.1-kg/cm²) maximum oil pressure is desirable. These recommendations follow a trend: 10-psi oil pressure per 1000-rpm maximum engine speed.

Which relief-valve-spring adjustments are made depend on many factors. Consider the following when making adjustments affecting maximum oil pressure:
- Oil weight (viscosity)
- Oil operating temperature
- Bearing clearances
- Oil-cooler size, design and plumbing

- Camshaft oiling—additional or external spray bar
- Crankshaft modifications— grooves or cross-drilling
- Cylinder-block oil-passage modifications
- Oil-pressure-gage accuracy

Even though 80—90 psi (5.7—6.4-kg/cm²) may seem to be the correct oil pressure for your wet-sump racing engine, multiple modifications, or not considering all of the factors affecting your engine's lubrication system, may result in exceeding oil-pump capacity. The resulting oil pressure may be only 60—70-psi (4.25—5-kg/cm²) at high engine speeds. However, many L-series engines have passed component inspection after numerous hours of hard running with this "low oil pressure."

EXTERNAL-PICKUP OIL PANS

Stock Pump—For wet-sump engine racing applications that retain the use of the conventional oil pump, one modification significantly improves oil pressure. Oil-pump performance is enhanced by not using the primary oil passages in the cylinder block. This requires using an oil pan

Oil pan modified for external pick-up hose. Dash-12 pick-up hose routes directly to stock oil pump. Dash-10 caps (arrows) seal discharge passage at oil pump and inlet at block. Later, a hose will be installed between these fittings.

with an external pickup and a new oil-pump cover. It also includes plugging the standard inlet and discharge passages either at the oil pump, the front cover or in the cylinder block.

The oil pan has the oil-pickup tube protruding from the right front corner of the pan sump. Weld a dash-12 male AN

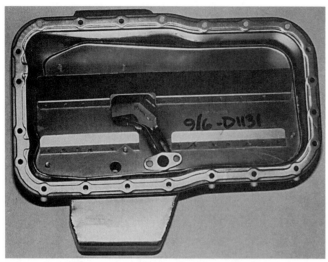
Nissan Motorsports' wet-sump oil pans are modified for racing. Full-length windage tray with built-in scrapers reduce oil drag on rotating bottom-end assembly and the resulting power loss.

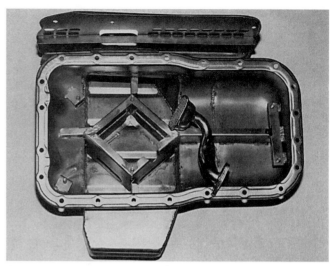
Sump area of wet-sump race pan modified to ensure that pickup is always immersed in oil. Sump capacity was increased from stock capacity of 4-3/4 to 7 qt.

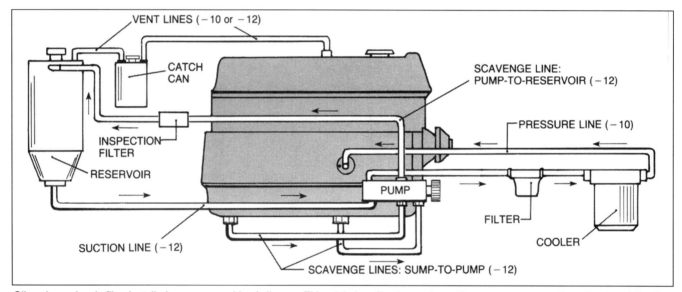
Oil cooler and main filter installed on pressure side of oil pump. This minimizes flow losses in suction, or scavenge, portion of system. Likewise, dash-12 lines are used for scavenging lines. Dash-10 lines are OK on pressure side. Screen-type inspection filter is used in pump-to-reservoir line to allow early detection of internal problems.

fitting to the oil-pan-pickup tube. The intake and discharge oil-pump passages are in the new pump cover, Nissan Motorsports part number 15015-GC002. Use a dash-12 male AN inlet fitting.

Connect the oil-pan-pickup tube to the pump inlet with a braided stainless-steel/neoprene dash-12 inlet hose. Because it is under negative pressure from the pump, anything less than braided steel hose may collapse and restrict oil flow, resulting in a severely damaged or destroyed engine. On the pressure side, the pump discharge should be routed through a dash-10 male AN fitting and hose to the remote oil-filter-inlet fitting.

Oil should flow from the remote oil filter to an oil cooler through a dash-10 hose. The size of the cooler depends on several factors: displacement and horsepower of the engine, location of the cooler and airflow through the unit. Remember that if the oil cooler is too large, it's easier to cover part of the cooler to raise the oil temperature than it is to make changes necessary to reduce oil temperature.

From the cooler, use a dash-10 hose to the fitting or adapter installed at the original oil-filter location. If either the fitting or an adapter is utilized, the pressure-relief

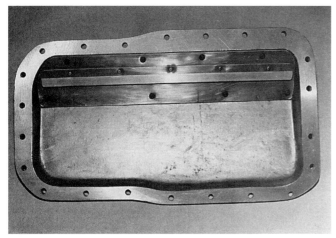

Crankshaft view of cast-aluminum ARE dry-sump pan shows built-in scraper. Strong pan gives support to block.

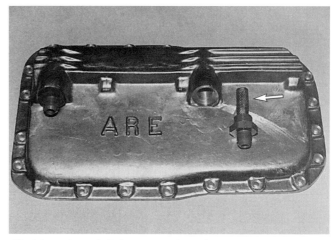

Oil pickups have built-in screens (arrow) that protect pump from ingesting debris in the event engine failure occurs. Two pickups are used; one at front center and one at rear of pan.

Dry-sump oil pump has very high capacity. Front two black sections of pump are suction or scavenge stages. Third, or smallest section is pressure stage for a turbocharger. Black section at rear is pressure stage for engine lubrication.

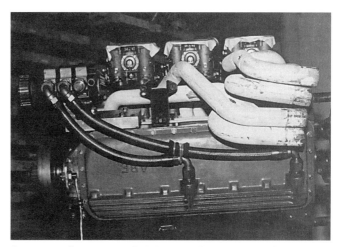

Dry-sump oil system on Bob Leitzinger's six-cylinder GTU engine utilizes a Weaver pump. Adjustment screw for pressure-relief valve is underneath pump at rear.

valve in the block must be removed and a solid plug installed in its place.

Wet-Sump Oil Pans—All wet-sump oil-system race engines need a special oil pan. First, reservoir capacity should be increased. To control the additional oil, the pan should be equipped with baffles and a windage tray to keep oil from sloshing around and away from the pickup.

Nissan Motorsports offers fabricated steel pans for L-series engines used with most chassis configurations. For four-cylinder engines, these pans have a 6.0—6.5-quart capacity. Modified pans for six-cylinder applications have a 7.0—7.5-quart capacity. All special pans have a full-length removable windage tray with two rows of scrapers to clean excess oil off of rotating components. In addition, each reservoir has an intricate baffle system that consists of a diamond-shaped baffle assembly with trap doors that open inward, toward the pickup.

During hard cornering or high-g stops, the trap doors on one side open to allow oil into the oil-pickup cup; trap doors on the opposite side close. This action ensures the diamond area is always full. Cornering forces on the oil are not a problem.

Dry-Sump Oil Pans—Nissan Motorsports also offers cast-aluminum dry-sump pans ideally suited for most four- and six-cylinder applications. These ARE pans have two dash-12 scavenge nipples. The cast-aluminum construction eliminates cracking and leak problems so common to fabricated steel pans.

When using a cast-aluminum oil pan—dry- or wet-sump—for off-road driving or other conditions with limited ground clearance, make sure the pan is protected by a skid plate or sump guard. Steel pans are less vulnerable, as they may only dent on impact, whereas the cast-aluminum pan may shatter like so much chinaware if it

Mounting for dry-sump pump must be rigid. Bracket shown is for Bob Leitzinger's Weaver pump.

Adapter fits cylinder block at original oil-filter location, Allowing use of remote oil filter and an oil cooler. If engine has an external-pickup pan and pump assembly, return fitting is used and outlet fitting is capped.

receives a direct hit from a rock or concrete parking-lot divider.

Cast-aluminum pans add additional rigidity to the bottom end of the cylinder block. There have been some incidents of oil-pan bolt heads snapping off on some six-cylinder, high-speed race engines. This apparently results from high loads between the rigid pan and the flexing engine block due to vibrations at high engine speeds. If bolt breakage occurs, install 6mm Allen-head bolts or increase bolt-hole size to 5/16 in. and install Allen-head bolts.

Dry-Sump Pump & Mounting Brackets—Most Nissan race engines require a three-stage dry-sump pump; two for *scavenging*—moving oil from the engine to the reservoir—and one for pressurizing the engine's lubrication system. If the engine is turbocharged, an additional scavenge stage can be for the turbo. Also, a larger pressure stage should be utilized. Dry sump-pump manufacturers such as Aviad, Peterson and Weaver can provide the correct pump for your engine application.

Many sources offer brackets for mounting the pump.

FOUR-CYLINDER CAMSHAFT SPRAY BARS

An add-on spray bar, similar in design to the one used in stock six-cylinder engines, is available to handle the rigors of high-performance operation and provide adequate cam-lobe lubrication.

This spray bar bolts to the inside of the valve cover. Oil is supplied by an external line routed up from the oil-pressure sender location in the side of the block. It can be used to supplement oil flow from the cam lobes. Or, it can be set up to replace direct oiling from the cam.

The spray bar has a series of holes that squirt oil directly onto each lobe where it contacts the rocker arm. The Nissan Motorsports spray bar is manufactured by Design Products.

These parts block off front-cover oil-pump mount and support pump/distributor-drive spindle when dry-sump-lubrication system is used.

CHAPTER TEN
Engine Electrics

Jim Goughary won the 1998 SCCA GT2 National Championship in an L28 powered 1995 300ZX. G. Hewitt photo.

DISTRIBUTOR
Street Applications—For most dual-purpose or street-performance engines, any stock Nissan/Datsun distributor is OK. Pre '74 or '75 conventional distributors are point-types. Afterward, electronic ignition was used. Some early point-type distributors use dual points. However, these dual-point distributors are not the performance types that permit longer point dwell for increased coil saturation. Instead, they actually have two separate systems that allow varied ignition timing for improved emissions control. Switches at the throttle linkage and transmission activate the separate systems. These distributors can be modified to improve high-speed performance.

Conventional single-point distributors provide relatively accurate ignition timing up to 6000 or 6500 rpm. However, these distributors must be modified to improve high-speed performance.

To modify a single-point distributor, the stock two-piece breaker plate must be secured or replaced with a one-piece plate. This eliminates the vacuum-advance function and reduces point bounce or flutter. On dual-point models, the secondary point-mounting plate must also be secured. This can be done by brazing the breaker-plate assembly solid so, in effect, it's a one-piece plate. Solid point-mounting plates are available from Nissan Motorsports. They install in place of the standard plate and utilize a single set of points. The solid plate eliminates the vacuum-advance function, but, as with the brazed stock plate, it mounts the points rigidly.

Distributor installation goes beyond slight modification to improve high-speed performance. Gene Crowe set up distributor to be driven off nose of cam. Mechanical fuel-injection-distribution unit is mounted in place of distributor.

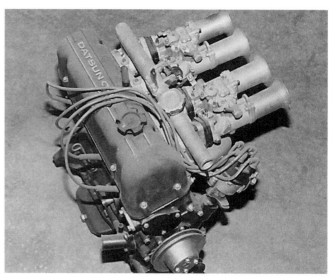
Four-cylinder race engine has potential redline of 8500 rpm. However, stock point-type ignition limits usable engine speed to about 7000 rpm. A $10,000 race engine with a $100 ignition system doesn't make sense!

SPARKPLUG CROSS-REFERENCE

Projected	Wide Gap		Regular	Fine Wire	
NGK BP(R)5ES Champion(R)N12Y	NGK BP(R)5ES-11	↑	B(R)5ES Champion(R)N8		Champion Gold Palladium
Champion(R)N10Y NGK BP(R)6ES Champion(R)N9Y	Champion(R)N10Y4 NGK BP(R)6ES-11	HOTTER	Champion(R)N7 B(R)6ES Champion(R)N6		
Champion(R)N8Y			Champion(R)N5 Champion(R)N4		Champion N4G
NGK BP(R)7ES	NGK BP(R)7ES-11		B(R)7ES	NGK B7EV	
Champion(R)N7Y				Bosch W4CS Champion N87	
Champion(R)N6Y NGK BP(R)8ES	NGK BP(R)8ES-11		Champion(R)N3 B(R)8ES Bosch W3CO	NGK B8EV Bosch W3CS	Champion N3G
NGK BP(R)9ES			Champion N2 NGK B9ES	Champion N2G NGK B9EV	
Bosch W2D Champion N60Y		COLDER ↓	Bosch W2CO Champion N60	Bosch W2CS Champion N84 Bosch W08CS NGK B10EV	Champion N59G
				Champion N82 Bosch W07CS	Champion N57G
				Champion N80	Champion N55G

Sparkplugs have 14mm threads, 0.750-in. reach and 13/16-in. hex for wrench. As listed from right to left, sparkplugs are interchangeable.

Nissan Motorsports also offers an ignition-points set with a much stronger spring arm. These points, which are for single-point distributors, further reduce point flutter or bounce. Mounted on a solid breaker plate, these points will give accurate ignition timing up to about 7000 rpm.

Inaccurate timing is the variation of ignition timing from one cylinder to the next. The performance of a point distributor can be further enhanced by adding an electronic control unit triggered by the points. This will increase the voltage delivered to the sparkplugs from the coil.

Nissan/Datsun electronic ignitions provide considerably more voltage to the sparkplugs. Also, they are not susceptible to point bounce because they utilize a *reluctor*—a pointed cam—and magnetic pick-up assembly to activate the primary ignition system. However, the performance

At left is European-model six-cylinder distributor. At right is SSS four-cylinder model. Both are good street-performance distributors.

Six-cylinder Mallory dual-point distributor has mechanical tachometer drive.

of an electronic distributor is limited at high engine speeds—over 7000 rpm—simply by how it is driven; by a gear at the crank and through the spindle to rotate the distributor shaft.

Spark accuracy with an electronic-type distributor is affected by clearance at the distributor-shaft and drive-spindle bushings, and backlash of the crank/spindle gears. Also, Nissan/Datsun electronic ignitions, like other conventional electronic-ignition systems, retard ignition timing at very high speeds.

Point-type and electronic distributors are available with a variety of advance curves. You can also tailor the advance curve by exchanging the springs and reluctor assemblies. Nissan Motorsports offers dual-point performance distributors with advance curves calibrated for performance applications. One is the SSS model for four-cylinder engines; the other is a European model for six-cylinder engines.

Race-Engine Applications—Accurate ignition timing at high engine speeds is not only important for performance, but also for engine reliability. If ignition timing is accurate at 7000 rpm with a conventional distributor, it will be completely unstable at 8000—9000 rpm. This unstable condition will lead to serious detonation and/or retarded ignition and a subsequent power loss. It should be noted that detonation on a 14:1 compression ratio race engine will cause instant engine failure. In some instances, inaccurate ignition timing is overlooked in diagnosing engine failures. Or, inaccurate timing is misdiagnosed as excessive spark advance.

L18 four-cylinder race engine built by Norm Balzer for John Koobation. Stock distributor is used to distribute secondary ignition, not to time it. Ignition is triggered at crank damper. Note electronic pickup behind damper under front cover (arrow).

To ensure accurate timing for all cylinders, the primary ignition system must be *triggered* at the crankshaft rather than by a points set or reluctor in the distributor.

John Ray of R.F.D. prepared 2.4-liter four-cylinder engine for Spencer Low's HDRA Class 7S Championship winning truck. The 16-gage wires connect magnetic pickup to an MSD control unit. The secondary coil wire is wrapped in a plastic sheath and routed to ignition coil in cab.

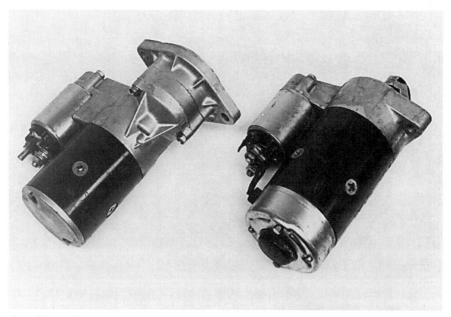

Early Nissan/Datsun L-series gear-reduction starter has become the most popular starter in racing. Nose housing can be substituted with a fabricated unit to adapt starter to most engines. Gear-reduction unit is at left; direct-drive at right. Photo by Tom Monroe.

This eliminates timing inaccuracies caused by the distributor-drive system, or backlash in the oil-pump/distributor-drive and -driven gears, windup in the drive spindle, and play in the distributor assembly that occurs at engine speeds over 7000 rpm.

In addition to accurate primary-ignition triggering, the available voltage from the secondary side of the ignition coil must be sufficient. At race speeds, the ignition-control unit and coil must recover rapidly to provide spark for the next sparkplug in the firing order. Consider the fact that at 8000 rpm the ignition coil for a six-cylinder engine must provide sparks 400 times a second!

Starters—In 1978, Nissan introduced a gear-reduction starter for the 280Z. This starter was improved over the years, but remained basically the same up to the 1983 280ZX. The gear-reduction starter has more than ample torque to start any L-series engine. It even has enough torque to crank an engine with a smaller-than-stock, 10-in.-diameter, low-inertia flywheel assembly.

As an example of how powerful the Nissan gear-reduction starter is, it is advertised by some aftermarket suppliers as a super-powerful, lightweight starter for use on everything from Formula Fords and small-block Chevys in Grand National stock cars, to Chrysler Hemis in dragsters.

The early-model Nissan gear-reduction starter has a two-piece nose housing. This allows the adaptation of the starter by fabricating a nose housing that matches that of the starter it replaces.

Nissan Motorsports offers a rebuilt and race-prepped gear-reduction starter that is the early type with the two-piece nose housing.

This Tilton-prepped starter has beefier hardware. Critical components are glued in place to reduce vibration-related failures.

Charging Systems—When considering a charging system, you must first determine whether your car needs one. For instance, if you race half-hour sprint, drag, short oval-track races or autocrosses, your car doesn't require a charging system. By comparison, if you compete in endurance races or rallies, a charging system with 70—80-amp output may be required. For racing applications requiring a charging system, Nissan has 35—80-amp alternators.

Standard-equipment Nissan/Datsun alternators produced after 1979 are all internal-circuit (IC) regulator types. These alternators are physically interchangeable with the earlier-type alternators. Conventional alternators are available with 35—60-amp output; IC alternators with 45—80-amp output. It is relatively easy to modify an early-style wiring harness for the later IC alternator.

Nissan/Datsun race engines are tough on alternators. Engine vibrations either shake the unit apart or, more commonly, shear the base-mounting bolts. Alternator internals can be glued in place. The alternator base and mount can be easily drilled to allow the use of a single 3/8- or 7/16-in. Grade-8 securing bolt. Use two tension brackets bolted together to secure the top of the alternator. To reduce alternator rotating speed, install an oversize pulley. These pulleys are either fabricated steel or aluminum machined from billet stock.

Six-cylinder L-series race engines that are revved to 8500 rpm are especially tough on alternators. For such applications, consider the following alternator-mounting system. This setup consists of a machined aluminum plate and modified Chevrolet engine mount. The top tension bracket is a length of aluminum hex stock with spherical rod ends and jam nuts at each end. At one end of the aluminum hex is a left-hand-thread 3/8-in. spherical rod end; a right-hand-thread 3/8-in. rod end is at the other.

The 3/8-in. rod ends have sleeved ball bores to make them compatible with the 8mm attaching bolts. Belt-tension adjustment is done by simply loosening the jam nuts, rotating the aluminum hex stock, then retightening the jam nuts. The combination of the rubber base mount and rod-end tension bracket provides a vibration-damped environment for the alternator, thereby greatly improving reliability.

Note: When using this alternator-mounting system, the alternator should be grounded with a ground strap. Otherwise, the rubber base mount and Teflon-lined rod ends may insulate the alternator, causing an open circuit.

Mounting is shown in photo at top.

Alternator-mounting setup is elaborate, but necessary for high-rpm endurance-racing engines. Ground wire between alternator and engine is required because base mount is rubber and spherical rod ends are Teflon-lined.

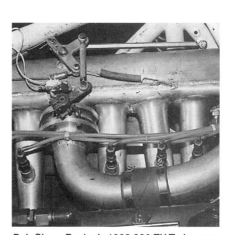

Bob Sharp Racing's 1983 280 ZX Turbo race engine uses ignition kill switch to protect car and driver against stuck throttle. Microswitch on throttle linkage must be closed when switch at throttle pedal is open, otherwise ignition is deactivated.

CHAPTER ELEVEN
Carburetion

Original BRE 510, car number-46, is fitted with pair of 44mm PHH Mikuni/Solex carburetors on individual-runner intake manifold.

View into throttle bore of Mikuni 50mm PHH carburetor shows main-venturi entry created by tapered air horn. Booster venturi is at center of throttle bore.

TWIN-CHOKE SIDEDRAFT CARBURETORS

The most popular high-performance carburetor for Nissan/Datsun engines is the twin-choke sidedraft type. Such carburetors are manufactured by Mikuni/Solex, Weber and Dellorto Construction. The principles of operation for all of these carburetors are basically the same. Also, each carburetor works best on individual-runner-type intake manifolds where there is one carburetor bore per cylinder. Manifold-runner size should match throttle-bore size. Finally, all three carburetors will bolt to the same manifold flange.

The following common features are found in Mikuni/Solex, Weber and Dellorto carburetors:

- Available in several throttle-bore sizes
- Selection of main venturis for sizing to engine-performance characteristics
- Different-length air horns, or velocity stacks, for runner-length tuning. This feature is often overlooked by engine tuners
- Large selection of main jets, air-correction jets, low-speed/idle jets and emulsion tubes/jet blocks. Emulsion tubes premix fuel with air before it reaches the main venturi
- Access to components listed above without carburetor disassembly
- Single float chamber per carburetor with single needle-and-seat assembly

Correctly tuned, these carburetors provide excellent performance. Most common complaints about carburetor performance result directly from the engine tuner not using all the available carburetor components to correct a performance deficiency. These carburetors are sensitive to jet sizing, fuel pressure and fuel level, bore size and venturi size in relationship to engine demand. These factors are easily controlled and adjusted, so they should not overwhelm the engine tuner.

Sidedraft carburetors provide excellent performance and trouble-free maintenance,

Carburetion 107

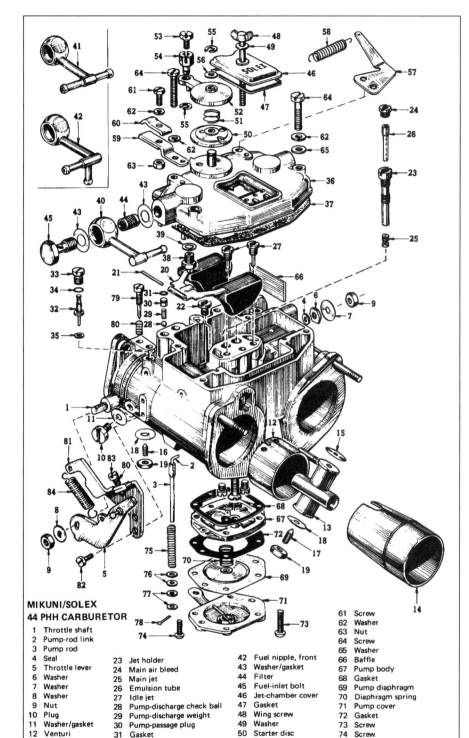

DCOE Weber main fuel jets, air-corrector jets, low-speed/idle jets and emulsion tubes are under jet cover at center of carburetor.

MIKUNI/SOLEX 44 PHH CARBURETOR

1. Throttle shaft
2. Pump-rod link
3. Pump rod
4. Seal
5. Throttle lever
6. Washer
7. Washer
8. Washer
9. Nut
10. Plug
11. Washer/gasket
12. Venturi
13. Booster venturi
14. Air funnel
15. Gasket
16. Setscrew
17. Setscrew
18. Washer
19. Nut
20. Float assy.
21. Float pin
22. Starter jet
23. Jet holder
24. Main air bleed
25. Main jet
26. Emulsion tube
27. Idle jet
28. Pump-discharge check ball
29. Pump-discharge weight
30. Pump-passage plug
31. Gasket
32. Pump nozzle
33. Setscrew
34. Gasket
35. Gasket
36. Float-chamber cover
37. Gasket
38. Needle-valve assembly
39. Washer/gasket
40. Fuel nipple, rear
41. Fuel nipple, center
42. Fuel nipple, front
43. Washer/gasket
44. Filter
45. Fuel-inlet bolt
46. Jet-chamber cover
47. Gasket
48. Wing screw
49. Washer
50. Starter disc
51. Starter spring
52. Starter cover
53. Capscrew
54. Cable collar
55. Snap ring
56. Washer
57. Spring bracket
58. Spring
59. Cable bracket
60. Cable clamp
61. Screw
62. Washer
63. Nut
64. Screw
65. Washer
66. Baffle
67. Pump body
68. Gasket
69. Pump diaphragm
70. Diaphragm spring
71. Pump cover
72. Gasket
73. Screw
74. Screw
75. Pump spring
76. Washer
77. Washer
78. Cotter pin
79. Idle-adjust needle
80. Spring
81. Setscrew bracket
82. Screw
83. Throttle-adjust screw
84. Throttle-return spring

if the engine tuner correctly selects or adjusts the following:
- Carburetor size
- Main-venturi size
- Air-horn length
- Emulsion tube
- Soft insulator mounts

Adjustments of each include:
- Fuel pressure
- Fuel level
- Idle-airflow synchronization or balance
- Throttle-opening synchronization

Final tuning includes selection of main jet
- Air-correction jet
- Low-speed jet
- Accelerator-pump stroke setting

SELECTING CARBURETOR SIZE

The following items should be considered when selecting sidedraft-carburetor size:
- Throttle-bore size
- Outer (main) venturi size
- Initial fuel and air jets

Major components of disassembled Mikuni 50 PHH carburetor. Weber, Dellorto and Mikuni carburetors are remarkably similar.

Three main venturis for 44 PHH Mikuni carburetor are interchangeable. From left to right venturi IDs are 32mm, 36mm and 40mm. Interchangeable venturis aid in tailoring carburetor size to specific applications.

Engine Dynamometer Test Log

Page No. _____ of _____
Barometric Pressure _____
Dry Bulb _____
Wet Bulb _____
Humidity _____
Vapor Pressure _____
Correction Factor _____
Engine Description _____

Owner _____
Date _____
Fuel Type _____
Exhaust System _____
Carburetor _____
Make/Model _____
Bore Size _____

RPM	GAGE	OBS BHP	CORR HP	FUEL FLOW	BSFC	TORQUE	MAIN JET	AIR JET	JET BLOCK	SPARK ADVANCE	OIL PRESSURE	EXHAUST TEMP	VENTURI SIZE

NOTES:

Engine-dyno sheet is typical of that engine builders use for recording data during engine testing. Data are compared after each run to determine whether gains were made with a change.

Carburetion

Air horns or velocity stacks can be used to tune engine. Opinions on how bell-shape entry affects performance differ, but effect of varying length of horns is common knowledge. For instance, longer air horns increase low-rpm torque.

At left is Mikuni 44 PHH; at right is 50 PHH Mikuni. Which size carburetor should be used depends on several factors. Engine size and rpm range are two major ones.

Road-Racing Carburetor Applications

SIX CYLINDER

IMSA 2.5 and 2.8 liter or SCCA Trans-AM 2.8 and 3.0 liter (Unlimited induction allowed)	Mikuni 50 PHH with 43 or 45mm venturi. Weber 48 DCOE with 42mm venturi or enlarged to 44mm.
SCCA GT-2 (formerly C/P) 280ZX	Weber 55 DCOE with 47mm venturi. Specified optional carburetor: Mikuni 44 PHH with required 36mm venturi.

FOUR CYLINDER

SCCA GT-3 (formerly GT-2) L18 and L20B	Mikuni 50 PHH with 43 or 45mm venturi, or Weber 48 DCOE with 42mm venturi or enlarged to 44mm. Weber 55 DCOE with 47mm venturi. Mikuni 44 PHH and Weber 45 DCOE have been used, but are not large enough.
SCCA GT-4 (formerly GT-3)	Mikuni 44 PHH or Weber DCOE with 40mm or no venturi, or Weber 45 or 48 DCOE with 40mm venturi.

Two views of same set of modified 50 PHH Mikunis: Carburetors in Keith Bowman's 200SX were extensively modified. Engine was built by Dave Weber of Malvern Racing. Air horns were machined from billet aluminum stock and bored to match taper of throttle bodies. Main venturis were removed to make "larger" throttle bores.

THROTTLE-BORE & VENTURI SIZE

Street-Performance & High-Performance Applications—

In applications where "big" carburetors are not required, there's a basic formula you can use to select carburetor throttle-bore and main-venturi sizes.

Only use the carburetor-sizing formula, page 111, as a guide for Nissan/Datsun L-series engines because it doesn't take into consideration specific engine modifications that will affect carburetor-size selection. However, it does consider the effects these modifications have on an engine's maximum usable rpm.

The two variables that must be determined for this formula are the volume of a single cylinder and the engine's maximum usable rpm or *horsepower redline*. The volume of a single cylinder is determined by dividing the total displacement of the engine in cubic centimeters (cc) by the number of cylinders. Maximum usable

110 How to Modify Your NISSAN/DATSUN OHC Engine

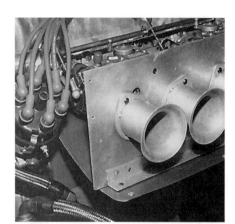

Author's IMSA GTU spec 2.8-liter race engine has Mikuni 50 PHH carburetors. Underside heat shield between exhaust headers and carburetors is stainless steel. Vertical plate sandwiched between air horns and carburetors is aluminum. Carburetor bodies are well insulated from exhaust heat.

Throttle cable from four-cylinder 510 engine is modified to adapt to L-series six-cylinder carburetion. Note banjo fitting with dash-4 male fitting (arrow) and starter-disc assemblies wired closed. Thin manifold-to-head flange resulted from flange being angle-milled to better align manifold runners with ports in head.

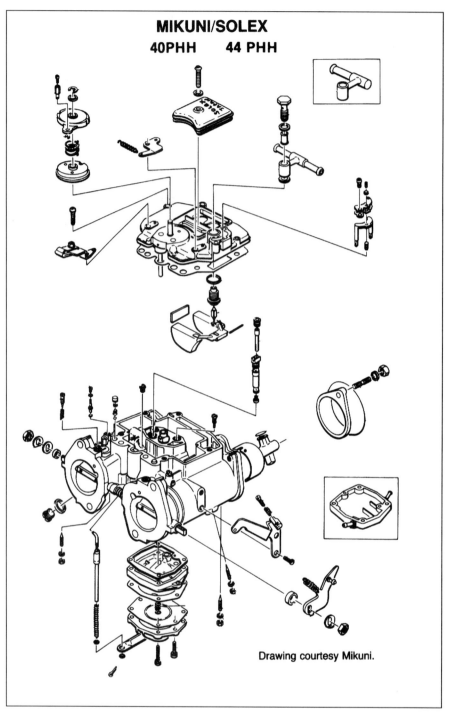

Drawing courtesy Mikuni.

engine rpm must be assumed. Maximum usable rpm depends on many factors that you must consider. If in doubt about maximum engine rpm, be conservative.

Racing Applications—Generally speaking, race engines are modified to produce maximum power at high rpm. The engine-rpm operating range where usable power is produced is typically narrow, but must be kept broad enough to produce sufficient power under most conditions, such as accelerating out of a low-speed corner. Maximum horsepower and high rpm means high induction-system airflow, which translates into the need for big-venturi and big throttle-bore carburetors.

Consider a 2.5-liter L-series six-cylinder racing engine used in an IMSA GTU car. This 152-cubic-inch displacement (CID) engine works well with three 48 or 55 DCOE Weber or 50 PHH Mikuni carburetors. However, by most standards, this would be considered an *over-carbureted engine*.

One complication in selecting carburetors for the L-series Nissan/Datsun

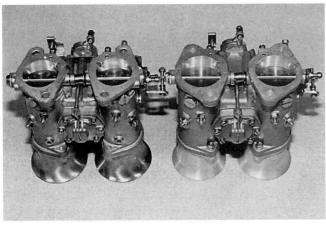
View of throttle-plate side of Mikuni carbs shows throttle-bore size difference. Unit at left is 44mm; carb at right is 50mm.

Partially disassembled late-style Mikuni 44 PHH carburetor shows all main-system components except accelerator pump.

engine comes in the form of restrictions from race sanctioning bodies. This generally involves a smaller-than-desirable carburetor main venturi. The result is reduced power output.

See list of popular carburetor-bore and -venturi sizes for Nissan/Datsun racing engines, page 109.

Throttle-bore size =

$\sqrt{\text{displacement per cylinder in cc} \times \text{maximum rpm} \div 1000} \times 0.82$

Main venturi size =

$\sqrt{\text{displacement per cylinder in cc} \times \text{maximum rpm} \div 1000} \times 0.65$

Example: Throttle-bore size
280Z, 2753cc engine, 6500-rpm maximum
Displacement per cylinder = 2753 ÷ 6 = 459cc
Throttle-bore size =
$\sqrt{459cc \times 6500 \div 1000} \times 0.82 = 44.8mm$.

Carburetor to match recommended throttle-bore size would be Mikuni 44 PHH or Weber 45 DCOE

Example: Main-venturi size
Same engine - same maximum rpm
Main-venturi size =
$\sqrt{459cc \times 6500 \div 1000} \times 0.65 = 35.5mm$.

Use 35 or 36mm main venturi.

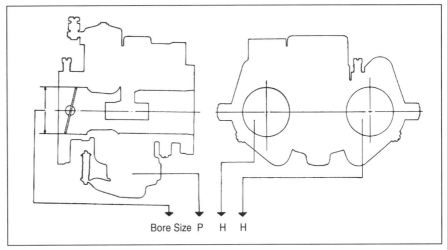
Mikuni/Solex PHH carburetors come in two models: 40PHH and 44PHH. P means accelerator pump and HH means horizontal-compound carburetor. Drawing courtesy Mikuni.

Initial Jet Sizing—For an initial main-jet size, start with one 0.04 the size of the main venturi. For example, if a 40mm main venturi is used, install a 1.60mm main jet, or 40 x 4 ÷ 100 = 1.60mm. For an initial air-correction-jet size, add 0.20 to the initial main-jet size. In the example, install a 1.80mm air jet, or 1.60mm + 0.20mm. = 1.80mm.

MIKUNI/SOLEX PHH CARBURETORS

Mikuni/Solex PHH sidedraft carburetors are available through Nissan Motorsports.

In addition to carburetors, Nissan Motorsports offers parts to service Mikuni carburetors, manifolds designed for racing, rubber insulators, linkage kits, throttle cables and heat shields. Nissan Motorsports was also the prime U.S. source for 50mm Mikuni carburetors.

To better explain the design, function and service of the Mikuni PHH carburetor, let's look at its five basic systems:
- Float
- Accelerator pump—one per carburetor
- Starter
- Pilot/low-speed
- Main

112 How to Modify Your NISSAN/DATSUN OHC Engine

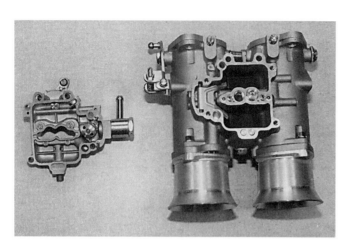

Mikuni fuel inlet and needle-and-seat assemblies are in float-chamber cover. Float and fulcrum are in main body. Float/fuel level can be measured through jet-block well. One jet-block assembly has been removed from carburetor.

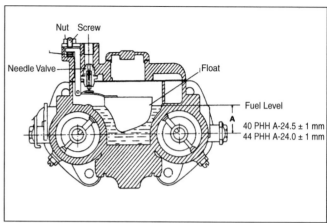

Fuel pressure should kept at 3.0—3.5 psi to maintain correct fuel level. If fuel pressure rises above this limit, fuel level will also rise. Drawing courtesy Mikuni.

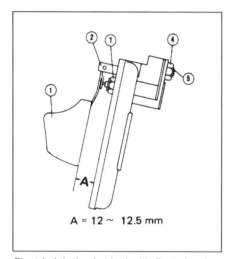

Float height is checked with float-chamber cover held as shown. Dimension A should be 12—12.5mm (0.47—0.47-in). Drawing courtesy Mikuni.

Late-style Mikunis have externally adjustable float/fuel height. Setscrew-and-locknut assembly (arrow) is positioned over needle-and-seat assembly. One turn changes fuel level 2mm (0.080 in.).

FLOAT SYSTEM

To operate, the float/fuel system must have fuel-inlet volume and pressure, a needle-and-seat assembly, and a float.

Mikuni/Solex PHH sidedraft carburetors are typically used on high-performance or racing engines with high fuel-flow requirements. Consequently, a fuel pump that provides enough fuel volume at a constant 3—3.55-psi pressure must be used. With such a fuel pump, it may be necessary to install a fuel-pressure regulator immediately upstream of the carburetors to regulate fuel pressure to 3—3.5 psi. This is because fuel pressure at the carburetor inlet has a significant effect on fuel/float level.

Fuel Level, Check & Adjust—To check fuel level without disassembling the carburetor, remove the jet cover—center top of the carburetor—and remove one jet block. The fuel level should be visible in the jet-block hole. Recommended fuel level is 0.826 in. (21mm) below the float-chamber top surface—horizontal surface at the top of the jet-block hole. This directly relates to

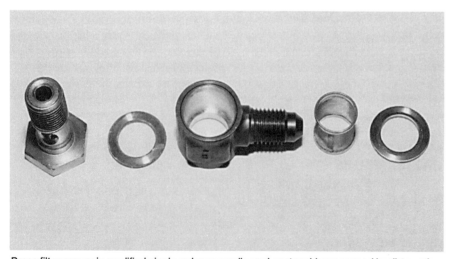

Brass filter screen in modified nipple reduces needle-and-seat problems caused by dirt or other contaminates. Nipple has dash-4 male fitting.

Carburetion 113

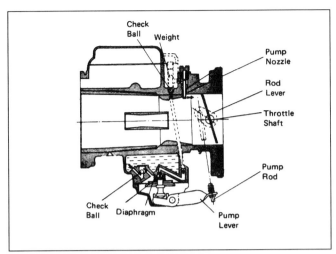

When throttle is opened rapidly from zero to 30% open, fuel is injected through accelerator-pump nozzle into each throttle bore as pump diaphragm is moved by pump pushrod and lever. Drawing courtesy Mikuni.

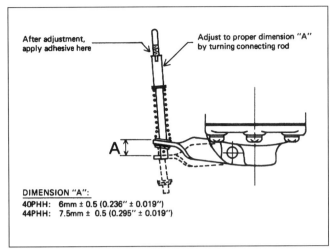

Amount of fuel Injected by accelerator pump can be changed by adjusting accelerator-pump pushrod. Drawing courtesy Mikuni.

Accelerator-pump assembly is centered under Mikuni PHH body. Pushrod activates lever arm to depress accelerator-pump diaphragm. Pushrod has three positions that can be used to adjust fuel flow.

Accelerator-pump diaphragm at right has spring attached to center of diaphragm to return diaphragm to its original position after activation to supply next shot of fuel.

Starter disc is rotated by cable that routes through cable clamp (arrow) of each of the two or three carburetors. Idle-mixture screws are in mounting flanges directly above throttle plates.

a fuel level 0.787 in. (20mm) above the horizontal center of the throttle bore on 50mm and early 44mm models and 0.940 in. (24mm) above the throttle bore on later carburetors.

Fuel-level adjustment methods for the Mikuni/Solex PHH vary. On early carburetors and all 50mm models, packing washers available in three different thicknesses install under the needle-and-seat assembly. These washers adjust needle-and-seat height and fuel level. The standard packing washer is 1mm thick, with 0.5- and 1.5mm-thick washers also available.

Changing washers adjusts the fuel level approximately 2mm per 0.5mm washer-thickness change. As an alternative, the float arms can be bent to change the fuel level.

Fuel level is externally adjustable on later-style carburetors. A vertical setscrew in the top of the carburetor body above the needle-and-seat assembly changes float-fulcrum height. Float/fuel level changes 0.080 in. (2mm) for each full turn of the setscrew.

Generally, most Mikuni carburetors have a 1.8mm (0.71 in.) needle-and-seat assembly. For racing applications, use a 2.0mm (0.080 in.) size. Mikuni America offers 1.2, 1.5, 1.8, 2.0 and 2.2mm needle-and-seat assemblies for both the standard early and later external float-adjustment carburetors.

ACCELERATOR PUMP

Mikuni PHH carburetors have a diaphragm-type accelerator pump on the underside of the body. The pump is activated by a lever/pushrod assembly attached to the throttle shaft between the two throttle bores. The pushrod has three adjustment positions to change pump stroke, thus increasing or reducing pump-discharge volume. There's

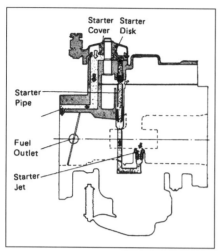

PHH starter system is used to richen air/fuel mixture for startup and cold-engine operation. Starter system should be deactivated is carburetor is used on race engine. Drawing courtesy Mikuni.

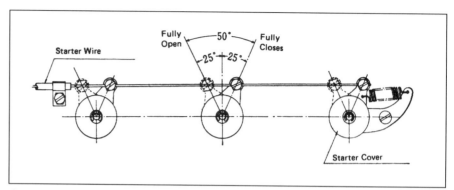

Choke cable rotates starters—three in this case—to activate. To deactivate, replace choke cable with length of wire, but with end of wire clamped in place of choke-cable conduit as shown in photo, page 110. Drawing courtesy Mikuni.

Low-speed/pilot jets (arrows) are adjacent to larger jet-block assemblies under jet cover.

Norm Balzer synchronizes airflow through Mikuni 50 PHH carburetors on L18 race engine prior to testing on Nissan Motorsports' dynamometer.

also an accelerator-pump nozzle that adjusts pump-discharge flow rate. This pump nozzle is in the top of the float-chamber body under the float cover. Different-size pump nozzles: 0.3—0.6mm in increments of 0.05mm are available from Mikuni America.

Correctly adjusting accelerator-pump flow rate and discharge volume will give a smooth transition during acceleration. A stumble or hesitation during rapid acceleration indicates a lean condition, while sluggish engine acceleration indicates a rich condition.

STARTER

The starter system of a Mikuni PHH carburetor is used instead of a choke valve for cold-engine starting.

On race cars, it is standard operating procedure to close the starter system securely or render it inoperative to ensure that it doesn't, for any reason, become operative at an inopportune time. Instead, the accelerator-pump charge is used to enrich the air/fuel mixture for cold startup.

When the starter is retained, the starter disc on top of the carburetor is rotated. This is generally done with a choke cable that rotates the starter disc. This opens both a fuel passage and an air passage. The opened air passage allows air to be drawn in at the starter cover. By leaving the throttle plate closed while cranking the engine, intake airflow draws fuel through the starter jet and up to the starter fuel passage. The starter fuel then mixes with air drawn in under the starter cover to enrich the air/fuel mixture and aid cold-engine starting and warmup.

The starter disc has three positions; off, partially open for warmup and fully open for cold starting. The disc rotates a total of 50° from closed to full open.

When the starter disc is rotated open, a vacuum—airflow—noise is evident around the starter cover as air is drawn in.

Remember, the throttle plate must

Carburetion

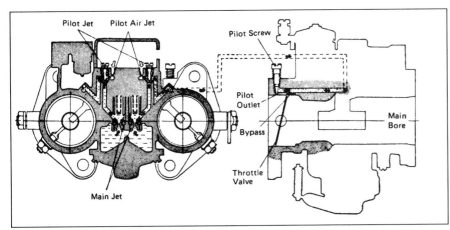

Low speed system of *homogeneous-type* Mikuni PHH carburetor consists of components such as pilot jet, pilot air jet, pilot outlet, pilot screw and bypass. Fuel is metered by jets and mixed with air metered by pilot air jets. Drawing courtesy Mikuni.

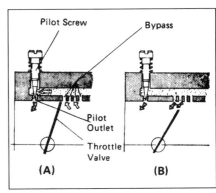

As throttle is opened and engine rpm increases, air/fuel mixture from pilot outlet eventually becomes insufficient. When this happens, air/fuel mixture begins to flow out bypass orifices in stages from left to right as valve continues to open. Drawing courtesy Mikuni.

be closed for the starter system to receive enough airflow to operate.

PILOT/LOW SPEED

The low-speed and idle circuit in the Mikuni PHH carburetor is referred to as the *pilot system*. The low-speed circuit functions up to approximately 15% throttle opening. The pilot fuel jets are available in different sizes. Also, there is an idle-mixture adjustment screw.

At idle, the mixture screw on top of the carburetor body above each throttle bore is used to adjust the air/fuel mixture. The pilot fuel-jet size also has some effect on air/fuel mixture, but is compensated for with idle-mixture screw adjustment.

Adjustment—As a general rule, the idle mixture screw should be 1.5 turns open to achieve the correct idle air/fuel mixture. More or less turns open may indicate the need to change the pilot fuel-jet size. Less than 1.5 turns open indicates the need for a leaner pilot fuel jet; a richer jet should be used if the screw is open more than 1.5 turns to achieve the correct air/fuel mixture. Each 0.5 turn from 1.5 turns will require one pilot-jet-size change.

Part of the idle adjustment is the synchronization of airflow through the multiple-carburetor system. This is done at the same time as adjusting idle speed and fuel mixture. A *Uni-Syn* carburetor-adjustment gage, Nissan part number EG167-00000, is

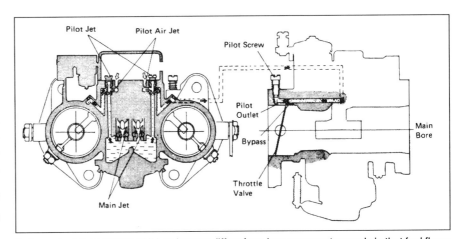

Independent-type Mikuni PHH carburetor differs from homogenous type only in that fuel flows directly to pilot jets rather than being routed through main jets. Otherwise, operation is identical. Drawing courtesy Mikuni.

the best tool for synchronizing carburetor airflow. Use an accurate tachometer to monitor and adjust idle speed. Idle air/fuel mixture should be adjusted to best-lean, or smooth idle with leanest fuel flow.

The low-speed air/fuel mixture—up to 15% throttle opening—is adjusted by changing the pilot fuel jet under the jet cover on top of the carburetor. The pilot fuel jets—one for each throttle bore—are alongside the jet-block assemblies. Pilot fuel jets are available in sizes from 25 to 110 with the most popular sizes being 45 to 70. Most Nissan/Datsun race engines use size 60 to 75, while more conservative ones use size 50 to 60 pilot jets.

Note: The lower the jet number, the leaner the air/fuel mixture.

MAIN SYSTEM

The main system operates when the throttle opens more than 15%. The main system consists of the main (fuel) jet, air (correction) jet, jet block (and bleed pipe—if so equipped).

Jets—Main jets are available in sizes from 50 to 210 (0.50—2.10mm) in increments of 5 (0.05mm), and from 210 to 250 (2.10—2.50mm) in increments of 10 (0.10mm). The air jet is available in sizes from 60 to 210 in increments of 5, and from 210 to 250 in increments of 10. Selecting main

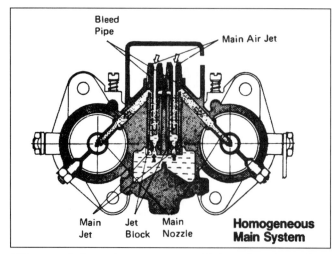 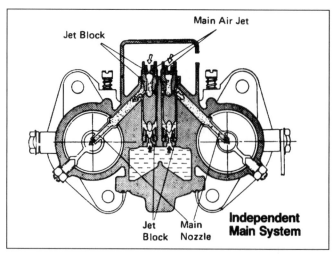

Differences between homogeneous and independent-type main systems: Homogeneous type includes bleed pipe between main jets and nozzles. Main difference between the two is mixture flow from main nozzle of independent type occurs earlier. Drawings courtesy Mikuni.

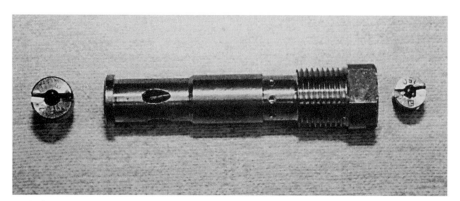

Main fuel jet is at left, or bottom of jet-block assembly. Air-correction jet installs in top or opposite end of jet block.

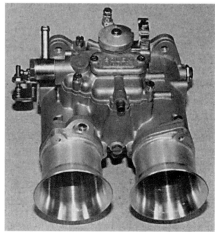

Main fuel-system components are under cover at center of 50 PHH Mikuni. Jet cover on very early 44 PHH model is same shape.

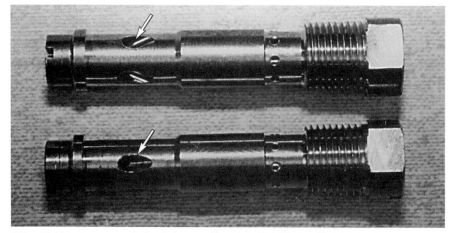

Mikuni jet blocks appear to be the same. However, bleed holes (arrows) under air jets and above fuel jets are different sizes. Variation in bleed-hole size affects transition mode.

and air jets isn't overly complicated. And, the two are relatively close in size to one another when correctly sized.

On most Nissan/Datsun racing engines, main-jet and air-jet sizes should be in the 160—200 range. Air/fuel mixture can be tailored to engine demand by changing the relationship between main- and air-jet sizes. Reducing main-jet size leans the air/fuel mixture across the entire rpm range; a larger main jet richens the mixture. The opposite occurs with the air jet. A larger air jet leans the air/fuel mixture at high engine speeds; a smaller one richens it.

The jet block (and bleed pipe—if so equipped)—also affect main- and air-jet size selection. The jet-block assembly is an emulsion tube with the main (fuel) jet attached to the bottom and the air (correction) jet attached to the top. The jet-

Carburetion

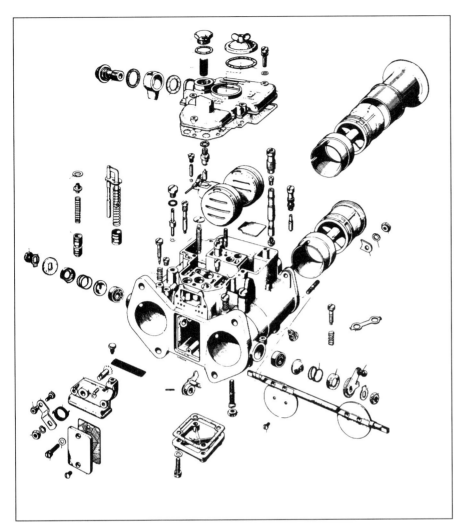

Weber DCOE is one of the most popular carburetors for racing. It has a high degree of parts interchangeability and is relatively easy to tune. Drawing courtesy Weber.

equipped with a fuel-flow meter or exhaust gas temperature meter (thermo-couple type). This is especially true on engines with unknown factors contributed by a special inlet system, exhaust system, camshaft design or compression ratio.

If a dynamometer is not available, you can still achieve good engine performance by using driver input, and by observing sparkplug, tailpipe and piston-top color.

WEBER DCOE CARBURETORS

Weber DCOE series sidedraft carburetors are available in throttle-bores ranging from 38 to 55mm. The most popular sizes are the 40, 45, 48, 50 and, most recently, 55mm. All Weber DCOE carburetors are nearly identical in construction and have interchangeable jets.

Weber DCOE carburetors comprise five systems:
- Fuel/float
- Idle/low-speed
- Accelerator
- Main
- Cold-start

FUEL/FLOAT SYSTEM

The fuel/float assembly is constructed of two thin-wall brass floats joined by arms that have a float-fulcrum pin and float tongue. The float assembly pivots on the fulcrum pin and operates the needle through the float tongue. Both solid and spring-loaded needle-and-seat assemblies are available in orifice sizes from 1.50 to 3.00mm.

To ensure correct adjustment and optimum performance, the related carburetor components must be in good working order. Before checking float level, remove the float cover and check for the presence of fuel in the float bodies. Also check the float tongue for wear at its contact point with the needle valve. Inspect the needle-and-seat assembly for contaminants, and make sure it is tightened into the float-chamber cover. The float assembly must swing freely on the fulcrum pin. Also, check the brass fuel-filter screen in the fuel inlet.

Fuel Level—Check and adjust float level while holding the float-chamber cover vertically with the fulcrum pin at the

block assemblies, one per venturi, are under the jet cover. Jet blocks equipped with bleed pipes are beneficial for part-throttle operation, but don't seem to be useful on racing engines that operate mainly at wide-open throttle.

The location and number of holes in the jet block—and bleed pipe—affect air/fuel mixture at different part-throttle openings and engine speeds. There are three jet blocks: OA, OB and 8. The standard is: 50mm carburetors for racing use the OB jet block, and 44mm PHH carbu-

retors for racing use the OA jet block. Bleed pipes aren't used in either application. Selective jet blocks with bleed pipes are available from Mikuni America for conditions requiring part-throttle tuning.

Tuning—When selecting jet sizes, start rich and work toward a lean mixture. A mistake on the lean side of the air/fuel mixture may be detrimental to your engine—melted pistons, for example. There's no better way to achieve optimum air/fuel mixture than tuning your engine on an engine dynamometer that's

118 How to Modify Your NISSAN/DATSUN OHC Engine

Weber carburetor has identification information stamped in top of float-chamber cover. TIPO—for type—is 55mm DCOE with accelerator pump.

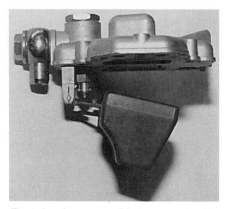

Float drop is measured between underside of float-chamber cover and closest part of float body. Float is not contacting the needle-valve assembly.

Idle speed/airflow is adjusted at throttle-stop screw (1). Idle-mixture adjusting screw (2) is immediately above each throttle plate (arrow).

Underside view of float-chamber cover shows float, float fulcrum, and needle-and-seat assembly. At top of photo is stock fuel-inlet banjo fitting.

Float level is measured when float-chamber cover is held in this position. Adjust float level by carefully bending tab LC. Distance between float and cover—with gasket in place—should be 0.275—0.315 in. (7—8mm) for L-series engines. Bend tab A to adjust float drop. Distance between float and cover should be 0.551—0.590 in. (14—15mm). Drawing courtesy Weber.

top. Float level is the distance from the float-chamber gasket surface with the gasket in place to the nearest part of the float body. On most Nissan/Datsun L-series engines, this distance should be 0.275—0.315 in. (7.0—8.0mm). The float tongue should be just resting against the needle valve. On a spring-loaded needle valve, take care to ensure that the spring is not compressed. To adjust the float level, carefully bend the float tongue near where it contacts the needle valve.

Another part of float-level adjustment is float drop. Check float drop by moving the float away from the float-chamber cover until the travel stop on the float arm contacts the float-chamber cover. The distance from the float-chamber cover—with its gasket installed—to the nearest part of the float body should be 0.551—0.590 in. (14—15mm). To adjust float drop, bend the float-drop stop tab.

Three other factors affect fuel level and should be considered. These are fuel volume, fuel pressure and carburetor-installation angle. If the carburetors are jetted correctly and the floats are set at the correct level, fuel volume will be adequate providing there's sufficient pressure. However, if there's not enough pressure, there won't be enough fuel volume. Fuel pressure should remain at 3.0—3.5 psi at all engine speeds.

As for mounting angle, most standard individual-runner manifolds installed on engines in the conventional position mount the carburetor at the correct angle, or 0—5° above horizontal. *Angle-milled* manifolds change the carburetor angle, thus affecting fuel level. Angle milling is done to direct the intake-manifold runners at raised ports in the cylinder head.

Mounting the carburetors rigidly with solid insulators causes engine vibrations to be transmitted to the carburetors, affecting needle-and-seat or float operation. Therefore, use rubber or other soft insulators to mount the carburetors.

IDLE & LOW-SPEED SYSTEMS

Engine idle speed, air/fuel mixture and airflow synchronization are all related. Consequently, they should be adjusted at the same time. Start by adjusting idle speed and airflow synchronization. Measure airflow with a Uni-Syn. Use an accurate tachometer to monitor and adjust engine speed. To achieve the desired engine idle speed and airflow synchronization, adjust

Carburetion 119

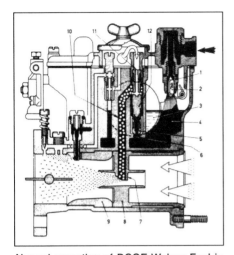

Normal operation of DCOE Weber: Fuel is drawn through nozzle (7) by vacuum signal generated at auxiliary venturi (8) and into throttle bore. Fuel is premixed with air drawn through air-correction jet (11). Drawing courtesy Weber.

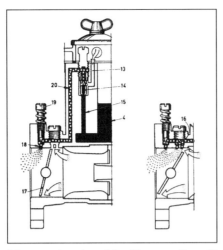

Idle-speed and part-throttle operation: At idle, fuel is drawn from bowl, is emulsified (mixed) with air drawn into ducts (13), then into throttle bore through feed hole (18) immediately downstream of throttle plate, or valve. At part throttle, additional fuel is drawn through progression holes (16). Drawing courtesy Weber.

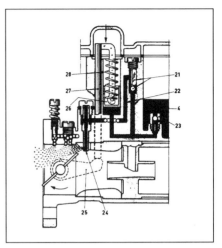

When throttle valve is opened quickly, accelerator-pump piston (26) moves down in bore, forcing fuel from pump jet (24), richening air/fuel mixture and giving rapid engine acceleration. Accelerator-pump discharge volume can be changed by changing rod length (27) and pump-jet size. Drawing courtesy Weber.

the throttle plates and/or linkage stop screws.

Idle-mixture adjusting screws are installed vertically above the throttle plates. Turning the mixture screws in—clockwise—leans the mixture; turning them out—counterclockwise—richens the mixture. As a general guide, the correct air/fuel mixture gives a smooth idle with the mixture screws one-half to one turn open. If more or fewer turns are needed to achieve smooth idle, larger or smaller idle fuel jets are needed. For example, if the mixture screws are open more than one turn, install idle fuel jets with larger bleed holes; if opened less than one-half turn, install smaller idle fuel jets.

Nissan/Datsun racing engines normally use a 0.55—0.65mm bleed-hole in the idle jets. Milder engines use a 0.45—0.55mm bleed hole.

Idle speed and airflow synchronization should be rechecked after mixture adjustment. The next step is to use the idle-speed adjustment screw to raise engine speed 300—400 rpm above idle speed. Turn the idle-mixture screws in or out. If low-speed mixture is correct, turning the mixture screws *in or out* will lower engine speed. If turning the mixture screw in rais-

Accelerator-pump piston-and-pushrod assembly is at top of carburetor (arrow). Linkage assembly is encased in carburetor body. Pump jets are over throttle bores under plugs at each side of pump. These plugs look like screw heads.

es engine speed, the mixture at low speed is too rich; if turning it out increases the engine speed, it's too lean.

The low-speed air/fuel mixture can be adjusted by changing the idle-jet fuel-feed hole size or the idle-jet air-bleed hole size. A larger air bleed will lean the mixture and a smaller one will richen the mixture. After selecting the proper idle-jet fuel-feed size

Auxiliary venturi is at center of each throttle bore. "Tail" or exit of auxiliary venturi extends into main venturi. Note drawing at upper left of this page. Air horn installed at left straightens airflow as it enters throttle bore.

for the idle air/fuel mixture, adjust the low-speed air/fuel mixture by changing idle-jet air-bleed hole size. The idle-jet air-bleed hole size is referred to as the *F-number* hole size.

ACCELERATOR SYSTEM

The Weber DCOE carburetor accelerator system uses a pushrod piston-type accelerator pump with fuel-flow-regulating

Safety wire (arrows) secures screws for main and auxiliary venturis. Bottom of float/fuel chamber is under cover at center of carburetor.

Main fuel jet, air-jet and emulsion-tube assemblies are under jet cover at center of float cover. This setup allows quick lot changes.

Jet well cover has been removed from underside of carburetor. Left main fuel jet is in place (arrow). Jet at right was removed. Low position of fuel jets in jet well reduces effects of side loads and bumps on fuel level.

components. Accelerator-pump discharge volume, discharge rate and flow rate are all adjustable.

Fuel for the pump well is drawn in from the bottom of the float chamber through the intake/exhaust valve in the bottom of the float chamber. The inlet-hole size of the intake/exhaust valve is not selective, but the exhaust-orifice size of the valve is. The exhaust-orifice function will be described later.

The throttle linkage-to-pump piston pushrod comes in four different lengths. Piston stroke can be varied from 10mm (59.5mm-long rod) to 18mm (67.5mm-long rod) to change total fuel volume discharged.

The volume of fuel flowing from the base of the pump well during discharge can be regulated by installing different size exhaust orifices in the intake/exhaust valve. Unrequired fuel recirculates through the exhaust orifice and back to the float chamber. Exhaust-orifice sizes are available from closed—no regulating—then from 0.35 to 1.00mm in 0.05mm increments. Most Nissan/Datsun L-series engines using individual-runner, intake manifolds use 0.45—0.55mm exhaust-orifice inlet/exhaust valves.

The stroke rate of the accelerator-pump piston can be adjusted with different-rate piston springs. Four different spring rates are available. Most L-series engines require the weak or medium-rate springs. The other two are heavy and extra-heavy springs.

The final regulating device is the pump jet, which regulates pump-discharge fuel flow into the throttle bore. The pump jet is available in eight sizes from 0.35—0.90mm. Most Nissan/Datsun L-series engines require 0.40—0.45mm pump jets. There are two pump jets per carburetor, one under a plug screw above each throttle bore.

MAIN SYSTEM
The Weber main system consists of the main (fuel) jet, air (correction) jet and emulsion tube. Two other components that are factors in the performance of the main system are the venturi and the auxiliary venturi.

How to select venturi size is described earlier in the general sidedraft-size section, page 111. Auxiliary-venturi size is also selective for the Weber DCOE. Sizes range from 3.50—4.50mm for the 40 and 42 DCOE, and 3.50, 4.50 and 5.00mm for 45 and 48 DCOE models.

Auxiliary venturis are available with or without air horns. Nissan/Datsun L-series engines generally require the 4.50mm auxiliary venturi. The auxiliary venturi is in the throttle bore, upstream of the main venturi and throttle plate.

The main jet, emulsion tube and air jet make up an assembly that installs in the center of the carburetor, under the jet cover. To do an initial main- and air-jet size selection, turn to the general sidedraft section, page 111. Main jets are available in 0.80—2.40mm sizes in 0.05 increments. Air jets are available in 0.80—2.40mm sizes, in 0.05mm. increments.

Most Nissan/Datsun engines use 1.4—2.10mm main and air jets, depending on main-venturi size. The main system begins to operate as the throttle opens beyond the low-speed stage. Fuel is then drawn from the float chamber into the jet well. This fuel is regulated through the main jet and mixes with air drawn through the air jet in the emulsion tube. The resulting air/fuel mixture passes to the auxiliary venturi where it blends with air flowing through the outer venturi in the throttle bore.

The location, size and number of holes in the emulsion tube have an affect on main-system performance. Eighteen (18) different-size emulsion tubes are available

Carburetion

to enrich or lean the air/fuel mixture at low and high engine speeds. Identifying numbers F2, F11, F15, F16 and F20 are the most common emulsion tubes for L-series engines.

STARTER SYSTEM

The starter system is used during cold-start and warmup conditions. All DCOE carburetors are equipped with starter systems except for three models. The 45 DCOE 12, 40 DCOE 20 and 22, and all DCOE 48 and 50 models have starter systems.

The starter is typically remotely activated with a *choke* cable from the driver compartment. The starter device has a fuel-enrichment position for cold starting and a partial-enrichment position for warmup. The throttle plate must remain closed for starter-system operation to obtain maximum manifold vacuum and so all air flows through the DCOE starter system.

For racing, the starter system should be made inoperative to ensure that it doesn't operate accidentally. The accelerator pump is then used to enrich the mixture during cold starts.

Calibration of the starter-system air/fuel mixture is possible by changing the starter jet. You'll find it under the float-chamber cover. A starter jet is essentially an emulsion tube with integral main and air jets. Main-jet size affects enrichment over the entire starter operating range. However, air-jet size affects it only during low-speed operation, such as during warmup.

SU CARBURETORS

The standard-equipment 1970—72 240Z & early SSS engines were equipped with Hitachi-built SU type sidedraft variable-venturi carburetors. Unlike the Mikuni or Weber carburetors discussed earlier, SUs are not the best choice for racing unless required by the rules. However, they are a good choice for street performance. They are relatively inexpensive, easy to maintain and are reasonably durable. Following is a description of the SUs, how they work, and how to maintain and set them up.

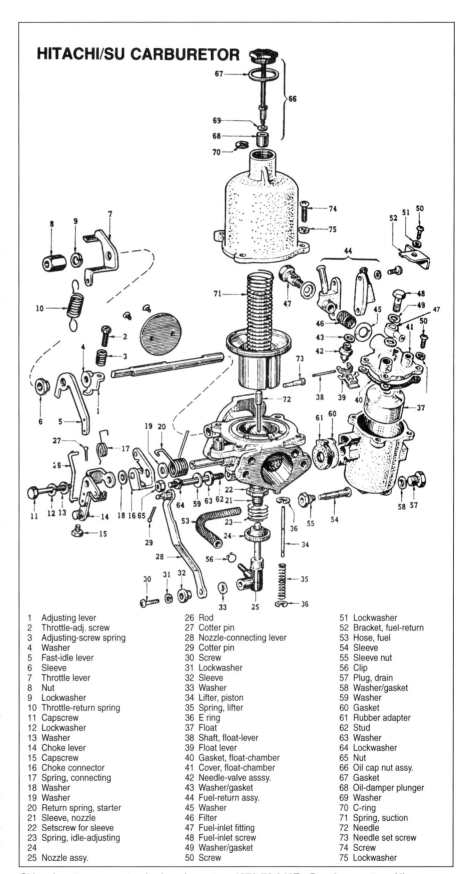

HITACHI/SU CARBURETOR

#	Part	#	Part	#	Part
1	Adjusting lever	26	Rod	51	Lockwasher
2	Throttle-adj. screw	27	Cotter pin	52	Bracket, fuel-return
3	Adjusting-screw spring	28	Nozzle-connecting lever	53	Hose, fuel
4	Washer	29	Cotter pin	54	Sleeve
5	Fast-idle lever	30	Screw	55	Sleeve nut
6	Sleeve	31	Lockwasher	56	Clip
7	Throttle lever	32	Sleeve	57	Plug, drain
8	Nut	33	Washer	58	Washer/gasket
9	Lockwasher	34	Lifter, piston	59	Washer
10	Throttle-return spring	35	Spring, lifter	60	Gasket
11	Capscrew	36	E ring	61	Rubber adapter
12	Lockwasher	37	Float	62	Stud
13	Washer	38	Shaft, float-lever	63	Washer
14	Choke lever	39	Float lever	64	Lockwasher
15	Capscrew	40	Gasket, float-chamber	65	Nut
16	Choke connector	41	Cover, float-chamber	66	Oil cap nut assy.
17	Spring, connecting	42	Needle-valve asssy.	67	Gasket
18	Washer	43	Washer/gasket	68	Oil-damper plunger
19	Washer	44	Fuel-return assy.	69	Washer
20	Return spring, starter	45	Washer	70	C-ring
21	Sleeve, nozzle	46	Filter	71	Spring, suction
22	Setscrew for sleeve	47	Fuel-inlet fitting	72	Needle
23	Spring, idle-adjusting	48	Fuel-inlet screw	73	Needle set screw
24		49	Washer/gasket	74	Screw
25	Nozzle assy.	50	Screw	75	Lockwasher

SU carburetors were standard equipment on 1970-72 240Zs. Drawing courtesy Nissan.

Datsun Alley, of Long Beach, California, prepared this 1970 240Z engine. The owner, Dick Berry, purchased European spec non-emission-control balance tube for intake manifold from Nissan Motorsports.

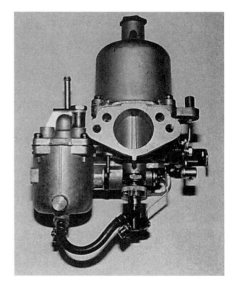

37mm (1.54n.) bore SSS SU: Suction piston in bore rises in dome-shape chamber as intake airflow increases to meter additional fuel.

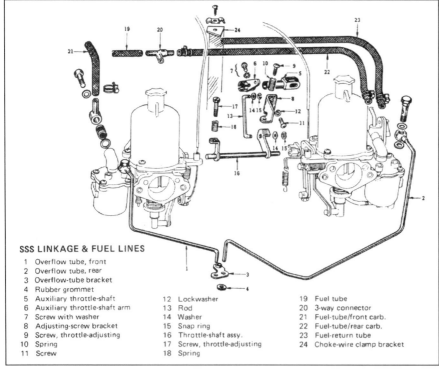

SSS LINKAGE & FUEL LINES
1. Overflow tube, front
2. Overflow tube, rear
3. Overflow-tube bracket
4. Rubber grommet
5. Auxiliary throttle-shaft
6. Auxiliary throttle-shaft arm
7. Screw with washer
8. Adjusting-screw bracket
9. Screw, throttle-adjusting
10. Spring
11. Screw
12. Lockwasher
13. Rod
14. Washer
15. Snap ring
16. Throttle-shaft assy.
17. Screw, throttle-adjusting
18. Spring
19. Fuel tube
20. 3-way connector
21. Fuel-tube/front carb.
22. Fuel-tube/rear carb.
23. Fuel-return tube
24. Choke-wire clamp bracket

Throttle linkage and fuel lines used with SSS SU setup. Drawing courtesy Nissan.

Fuel hose connects fuel nozzle with remote fuel chamber at side of throttle body.

Basic Construction & Function—

The throttle body has a throttle-linkage-operated throttle plate and a vertical sliding piston that varies the area of the venturi opening. Positioned above the throttle body is the suction chamber in which the sliding suction-piston assembly operates. Suction-piston movement is damped with oil.

When the throttle plate opens, intake airflow acts on the top side of the suction piston through a port in the suction chamber. As airflow draws the piston assembly up, the metering needle attached to its underside raises out of the fuel-supply nozzle in the front of the throttle body. The tapered needle allows more fuel to be drawn from the supply nozzle

Carburetion

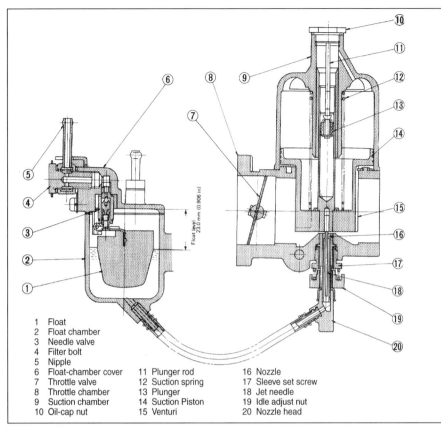

1. Float
2. Float chamber
3. Needle valve
4. Filter bolt
5. Nipple
6. Float-chamber cover
7. Throttle valve
8. Throttle chamber
9. Suction chamber
10. Oil-cap nut
11. Plunger rod
12. Suction spring
13. Plunger
14. Suction Piston
15. Venturi
16. Nozzle
17. Sleeve set screw
18. Jet needle
19. Idle adjust nut
20. Nozzle head

Suction piston rises and falls in suction chamber to change area between jet needle and fuel nozzle as engine fuel demands change based on airflow through venturi. Drawing courtesy Nissan.

Knurled knob at base of fuel nozzle (arrow) is adjustment for fuel-mixture ratio. Changing nozzle position affects mixture throughout operating range.

according to airflow as the piston-and-needle assembly rises.

The remote fuel/float chamber is mounted on the side of the throttle body. The needle-and-seat assembly is mounted in the throttle fuel-chamber lid. A rubber hose connects the fuel chamber to the fuel-supply nozzle. The fuel level in the chamber, when properly adjusted, assumes a level immediately below the top surface of the supply nozzle.

IDLE SPEED, IDLE AIRFLOW SYNCHRONIZATION & IDLE AIR/FUEL MIXTURE

When two carburetors are used, you must adjust both at the same time. With the engine warmed to operating temperature and the choke off, remove the air-cleaner assembly.

You'll need a Uni-Syn to measure airflow through the carburetors at idle. Idle speed and airflow through each carburetor is adjusted—synchronized—in the same operation.

First, adjust fuel flow at idle. While the engine is idling, reach in through one carburetor throttle body and raise the suction piston assembly about 0.5 in. This makes the carburetor inoperative. The engine should run as though it were operating on only two cylinders; three if it's a six-cylinder engine. If the engine dies, the air/fuel mixture is probably lean. If it runs slightly rough, the mixture is probably rich.

Fuel-flow mixture is adjusted by either raising (leaning) or lowering (enriching) the nozzle-height adjusting nut on the underside of the carburetor. After adjusting one carburetor, repeat the operation on the other. When you've finished adjusting the fuel-flow mixture, you may need to readjust engine idle speed and recheck airflow synchronization.

INSPECTION & SERVICE OF SUCTION-PISTON/CHAMBER ASSEMBLY & NOZZLE

The suction-piston damping-oil level requires periodic inspection at about every 2—3 months or 2000 miles. To do this, remove the oil-cap nut from the top of the suction chamber and the plunger and rod that are attached to it. The plunger rod is graduated so the oil level in the piston cavity can be measured. When refilling the suction chamber, it's generally best to use lightweight engine oil—about 20W. However, it's OK to experiment with lighter oil—try ATF or heavier weight oil according to temperature conditions and desirable engine characteristics. Use lighter oil if temperatures are low; heavier when temperatures are high.

Check the suction piston for free movement by first removing the oil cap nut/plunger-and-rod assembly. This will eliminate the effect of damping oil on piston-assembly movement. Lift the piston with a finger in through the throttle body, then release it and let the piston drop. It should move up and down freely. When released, there should be a slight "thunk" as the piston drops onto the venturi floor. If the piston doesn't move freely, the jet needle may be bent and rubbing on the inner diameter of the nozzle. Or, piston movement may be inhibited by dirt in

90-SERIES SU CARBURETOR NEEDLES
(Diameters given are from shoulder to tip end of noodles.)

	16354-2211 SSS	16554-H2316 GX	16354-15810	16354-A8710	16354-E3210	16354-H6000	16354-A7710	16354-16710	16354-10400	16354-12210	16354-14610
Number on Needle	M66	M83	M43	M76	M73	M87	M70	M6	M26	M15	M39
0.00	0.089	0.088	0.091	0.088	0.0885	0.0875	0.089	0.0895	0.090	0.090	0.090
0.100	0.087	0.0875	0.0875	0.087	0.087	0.086	0.088	0.0875	0.088	0.088	0.088
0.400	0.081	0.0815	0.081	0.081	0.081	0.080	0.082	0.082	0.0825	0.0825	0.082
0.700	0.074	0.0755	0.075	0.074	0.075	0.0735	0.074	0.076	0.078	0.0775	0.076
1.000	0.0685	0.070	0.0695	0.069	0.070	0.0685	0.068	0.073	0.072	0.072	0.071
1.400 down	0.064	0.0625	0.0665	0.0635	0.065	0.063	0.060	0.070	0.064	0.0635	0.062

Column at left is distance from shoulder of needle to tip. Numbers in columns at right are needle diameters at listed distances from shoulder for each needle.

the piston-and-chamber assembly.

To clean the piston-and-chamber assembly, remove the four screws that secure the chamber. Remove the chamber and *carefully* lift out the piston assembly. Don't bump or bend the jet needle that projects from the piston underside. Clean the inside of the chamber, piston and piston area of the throttle chamber.

Inspect the jet needle for straightness and correct positioning in the piston. If the needle is worn, it is probably bent so it rubbed against the nozzle. If this is the case, replace the needle. The shoulder of the needle should be flush with the underside of the piston. To reposition or replace the needle, loosen the needle securing screw in the bottom side of the piston. Handle the needle with care.

Inspect the jet-needle nozzle in the bottom of the throttle-chamber venturi. It should be perfectly round. If the nozzle is worn, replace it. If service to the nozzle is required, the carburetor should be removed from the intake manifold if it's still attached. Remove the nozzle from the underside of the carburetor. First remove the starter link from the nozzle, then the fuel hose and fuel-mixture adjusting nut and spring, and finally the nozzle.

Don't remove the nozzle sleeve. It is very difficult to re-index it and the nozzle

Jet nozzle is visible at base of throttle chamber. Jet hole must be perfectly round. A bent needle may cause wear at jet-nozzle bore.

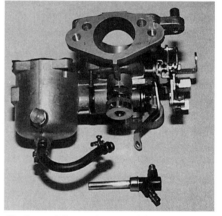

Choke-enrichment lever and fuel hose must be removed prior to removing nozzle assembly. Nozzle assembly and jet needle should be replaced it one or the other is worn or damaged.

so the sleeve is perfectly parallel to the jet needle. Reassemble the nozzle in reverse order and adjust the fuel-mixture adjusting nut three turns down from the top or to its original position for initial engine startup.

During reassembly, insert the piston into its throttle-chamber position. Index it in the locating tab and install the spring. Reinstall the suction chamber and check for up-and-down piston movement. Check that the needle does not contact the nozzle during piston movement. Refill and check the damping-oil level before starting the engine.

INSPECTION & ADJUSTMENT OF FLOAT LEVEL

The fuel/float chamber is mounted externally to the side of the throttle body. Fuel passes through the fuel hose from the bottom of the fuel bowl to the base of the nozzle where it settles in the nozzle at the same level as fuel in the fuel chamber.

One method of inspecting the float

Carburetion

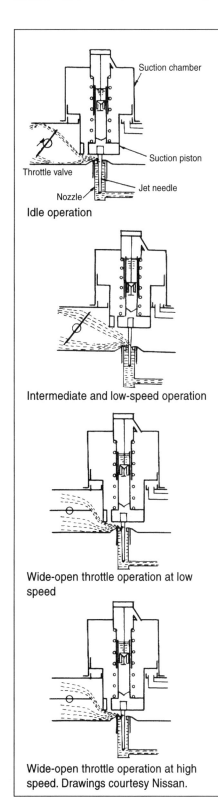

Idle operation

Intermediate and low-speed operation

Wide-open throttle operation at low speed

Wide-open throttle operation at high speed. Drawings courtesy Nissan.

Invert float-chamber cover as shown to measure distance from cover to float arm. This distance relates directly to fuel level in chamber and at jet nozzle.

Brass float assembly is found in most SSS SU carburetors. Needle-and-seat assembly is mounted in float cover. Float lever and fulcrum pin is at bottom of photo.

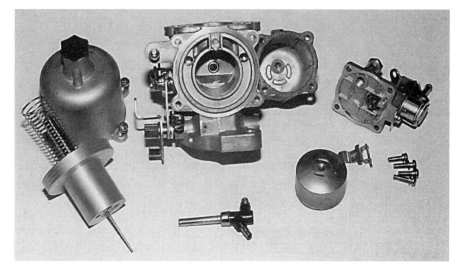

Reconditioned SU Carburetor ready to be reassembled. SU carburetor is very simple and easy to service and maintain.

level is to check the fuel level in the jet nozzle with the suction chamber and piston assembly removed. The fuel level should be just below the top of the nozzle.

Remove the fuel-chamber cover to adjust the float level. Invert the fuel-chamber cover and measure the distance from the closest point of the float arm to the cover. This distance should be 14—15mm (0.55—0.59 in.). Bend the float lever to obtain the correct distance. Also check the condition of the float in the float chamber. Replace the float if there's fuel in it.

The needle-and-seat assembly is in the underside of the fuel-chamber cover. Remove the float-lever fulcrum pin to gain access to the needle and seat. Correct needle-and-seat and float operation requires that the fuel-supply pressure be maintained at 3—3.5 psi. In some applications, it may be necessary to mount the carburetors to the intake manifold with soft rubber insulators. These soft insulators help to isolate the carburetors from harsh engine vibrations that can adversely affect carburetor operation. If rubber insulators are not used, install Bakelite isolators to isolate engine heat from the carburetors. This will reduce heating of the fuel in the float bowl and help to prevent vapor lock.

CHAPTER TWELVE
Exhaust

Jim Conner charges through Mexican desert during running of 1977 SCORE Baja 1000. Exhaust systems for off-road vehicles should be designed similar to that for high-performance street machines for bottom-end torque.

Now that you've built your engine and have the induction system in place, you'll have to make provisions for getting the burned air/fuel mixture out of the engine. Also, because top performance is a prime consideration, you should route the exhaust from the engine in a way that will enhance its power. This is done differently for different engine applications. Let's look at some typical applications.

STREET EXHAUST

Mild Street—For any stock or slightly modified engine, the original equipment exhaust manifold is suitable. But, to improve performance, the exhaust pipe and muffler should be replaced. On most four-cylinder engines, a 2-1/4-in. pipe should be used. A 2-1/2-in. pipe works best with a six-cylinder L-series engine.

The most popular performance muffler is the reverse-flow Corvair type that's available from most performance exhaust-system manufacturers. These are commonly called *turbo mufflers* because they were designed for use on production turbocharged Corvairs. This muffler type is free-flowing, relatively compact and reasonably light. For four-cylinder applications, most Corvair-type mufflers are OK. For six-cylinder applications, muffler-inlet and outlet size should be increased to 2-1/2 in. The pipe should not neck down inside the muffler.

Finally, some low-cost Corvair-type mufflers are short-lived and can cause annoying resonance. So, spend a little more for a higher-quality muffler. Not all turbo mufflers are created equal.

Wild Street—If you've modified your street engine, a good exhaust header will enhance engine performance. A radical street-performance engine should have characteristics similar to those of an off-road-racing engine. More attention should be paid to obtaining bottom-end—low-rpm torque and a wide power band than obtaining maximum horsepower, which can only be gained at high rpm. With this in mind, most street-performance engines should have 1-1/2-or 1-5/8-in.-diameter primary pipes 34—36in. long. Which pipe diameter you should use depends on your engine's power band, maximum rpm and displacement.

As an example, a 1600cc four-cylinder or 2400cc six-cylinder engine with moderate modifications and a maximum engine speed of 5500 rpm will require 1-1/2-in. primary pipes. By comparison, a 2000cc four or 2800cc six with many high-performance modifications may need 1-5/8-in. primary pipes with a 2-1/2-in. collector and exhaust pipe for maximum performance.

To obtain optimum performance, the primary header pipes should be the same length or within 2 in. of each other. On four-cylinder engines, the primaries should come together in one collector. Sixes should use two collectors. Number 1, 2 and 3 primaries go into one collector and number 4, 5 and 6 primaries go into the other. The two collectors are then connected by a Y-pipe.

Street Turbocharging—Use a cast-iron exhaust manifold with turbocharged engines. It will give better performance for

Tom Wyatt III—Turbo Tom's, Atlanta, Georgia—enters turn 1 at Road Atlanta during the "Walter Mitty." Powered by turbocharged L20B, 510 four-door car was nine seconds faster than street-radial shod 930 Turbo Porsche and three seconds faster than 427 Cobra! Photo courtesy Turbo Tom's.

L20B in car shown at left. Custom installation includes T04 Roto-Master turbocharger, Turbonetics wastegate and 4150 Holley carburetor. Engine develops 275 HP at 21-psi boost for drag racing and 250 HP at 17 psi for road racing. Note 4-in.-diameter turbine-outlet pipe. Such an installation requires custom-fabricated manifolds. Photo courtesy Turbo Tom's.

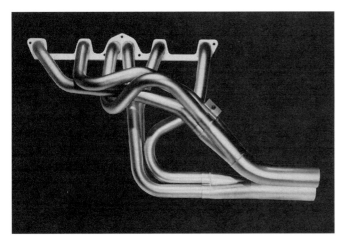

Cyclone header for 1970—74 Z-car is equipped with air-injection fittings. Header is made for square exhaust port heads. It has 22—24-in.-long, 1-5/8-in.-diameter primary pipes and 2-1/4-in.-diameter collectors.

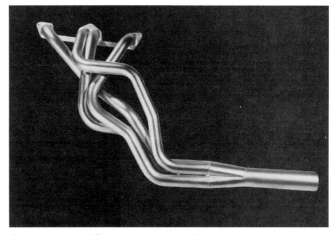

Collector outlet of Cyclone header uses flanges for attaching exhaust pipe. Setup is better for street use because of improved sealing.

the street and will be much more durable than tubing-type headers.

RACING EXHAUST

Exhaust-header design can significantly affect engine performance. Primary-pipe diameter and length have a marked effect on the power band and peak engine power.

To see how pipe diameter can affect engine power, note that a 1-5/8-in. pipe has 17.5% more cross-sectional area than a 1-1/2-in.-diameter pipe. But, a 1-3/4-in. pipe has 36% more area than a 1-1/2-in. pipe. Consequently, even though tubing size increased by 8.3% and 16.7%, areas increased by 17.5% and 36%, respectively! Such changes in cross-sectional area and resulting flow in the exhaust system greatly affect back pressure. This back pressure, or its absence, also affects airflow into, within and out of the combustion chamber. This, in turn, affects airflow at various engine speeds or cylinder

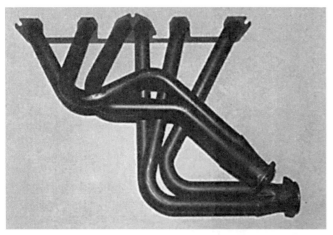
Clifford Research header fits Z-car with round exhaust ports. Primary pipes are 27—29-in. long and 1-1/2-in. round. A 2-1/2-in. collector is used. There are no provisions for air-injection fittings.

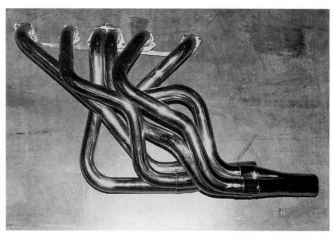
Nissan Motorsports' 1970—78 Z-car header 99996-E1142 is for use with square exhaust ports. Header used 32-in.-long 1-3/4-in.-diameter primaries and 2-1/2-in.-diameter collector.

Motorsports' header is constructed from mild steel, then is chrome-plated. Chrome plating reduces maintenance. Header flange matches square exhaust ports. Header should not be used with round-port heads and vice versa.

Mocking up exhaust header for Bob Sharp's SCCA GT-1 L24 Turbo engine: Partially completed headers are shown at upper right on following page. Photo by Ron Fournier.

pumping speeds.

Too much back pressure will limit high-rpm power because the exhaust system won't be able to handle the high flow rate of the exhaust gases. However, if pipe diameter—area—is too large, flow velocity and the resulting scavenging of the exhaust gases and the amount of air/fuel mixture drawn into the combustion chamber will be reduced at a given rpm.

To achieve the best all-around race performance, a header with 32-in.-long, 1-5/8-in.-diameter primary pipes is often the best. Such a setup gives the best power curve, that is, power over a wide rpm range.

So, by tailoring certain exhaust-header dimensions, power can be enhanced at certain engine speeds. Extremes include the "stump-puller" with good low-rpm torque, or the peaky engine with good high-rpm power, that has trouble accelerating from a dead stop. A lot of power in a narrow range or band at high rpm is characteristic of a peaky engine.

A last thought before we look at specific race applications: Where it mounts to the head, each primary pipe should match or be slightly larger than the cylinder-head port. If the pipe is smaller, port-to-primary pipe flow will be inhibited. If the match is perfect, there will be no restriction. Also, if the pipe is too large, there should be no restriction. In fact, a too-large primary pipe at the port-to-primary pipe transition may enhance power.

With a primary pipe that's larger than the port, flow in the opposite direction may be inhibited, reducing the contamination of the incoming air/fuel mixture by exhaust gas. This occurs when exhaust flows back into the combustion chamber with certain valve overlaps and induction setups. In fact, an *antireversion* (AR) header—one designed to inhibit flow in the reverse direction—may be the best choice. Unfortunately, AR headers are not readily available for four- or six-cylinder L-series engines. So, if you want a set, they'll have to be custom-built.

Determining exactly what type of exhaust header works best requires that your engine be dyno-tested. This is just as it would be for different cam and cylinder-head port combinations.

Off-Road Racing—For a broad torque curve, an off-road racing truck four-cylinder engine works best with 36-in.-long, 1-5/8-in. primaries that meet in a single

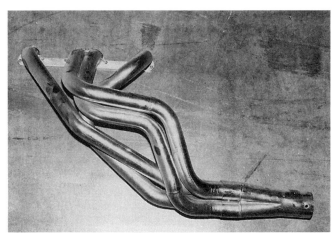
Nissan Motorsports header 99996-N8000 Is for 1977—79 200SX with stock round exhaust-port head. Header has 1-5/8-in.-diameter primaries and 2-1/2-in.-diameter collector.

Custom-built, stainless-steel header is the most expensive variety. However, if cracks are avoided on a turbo race engine, the additional cost is worth it. Photo by Ron Fournier.

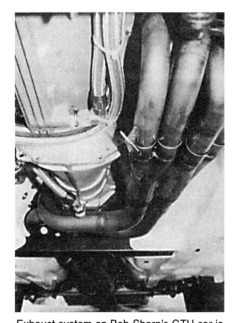
Exhaust system on Bob Sharp's GTU car is nicely tucked underneath beside transmission and in drive-shaft tunnel. Photo by Ron Fournier.

John Olsen's L16 powered GT4 class Nissan 200SX. G. Hewitt photo.

collector. The collector then empties into a 2-1/2-in. pipe.

Road Racing—At the other extreme, a 2000cc or larger four-cylinder engine or 2800cc or larger six-cylinder road-racing engine should have 1-3/4in.-diameter, 32-in.-long primaries. The larger-diameter primaries handle additional exhaust-flow volume resulting from high-rpm operation. Shorter primaries enhance high-rpm power.

On the six-cylinder, six primaries empty into two collectors; numbers 1, 2 and 3 into one and numbers 4, 5 and 6 into the other. The collectors then empty into a pair of 2-1/4 or 2-1/2-in.-diameter exhaust pipes at least 72 in. long.

On 2000cc or larger four-cylinder engines, a 3-in.-diameter exhaust pipe at least 72 in. long should be used.

Turbocharging—I've said nothing about exhaust-header material until now. That's because mild-steel tubing will do for naturally aspirated racing engines. It won't, however, work for a turbocharged engine because of the danger of cracking. Mild steel can't handle the stresses caused by the high exhaust temperatures resulting from turbocharging. One crack means lost boost, power and possibly a race. To reduce the possibility of this happening, use stainless-steel tubing with turbocharged engines when using tube-type exhaust headers.

Yes, there's an engine under all that plumbing. Engine on Electramotive dynamometer is from 280ZX shown on previous page. Note excellent workmanship on custom-fabricated exhaust system.

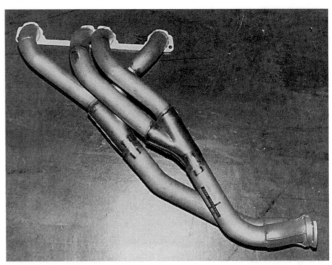

Nismo Nissan Japan—factory racing header fits optional FIA Group-2 race head. Header is designed for 1977—80 HL510. Tri-Y design is as unique as stainless-steel tubing. Very expensive header is no longer available (NLA).

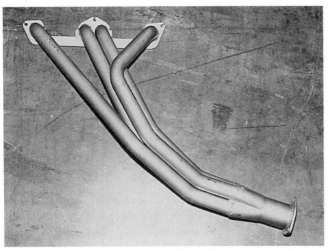

Strange-appearing header Nissan 99996-T1022 fits L20B in four-wheel-drive truck. Extra-long primaries are for low- and mid-range power. Header is no longer available (NLA).

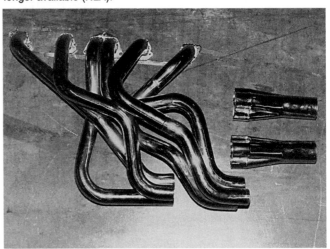

Nissan Motorsports' race headers have integral collectors for improved packaging, and installation and removal. Header is no longer available (NLA).

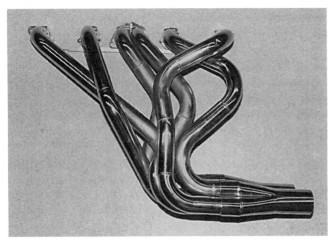

Electramotive race header is for 280 ZX equipped with recirculating-ball-type steering gear. The 1-3/4-in.-diameter primary pipes are 32-in. long. Collector diameter is 2-1/2 in., although many Z-car race engines work best with 2-1/4-in. exhaust pipes. Header is no longer available (NLA).

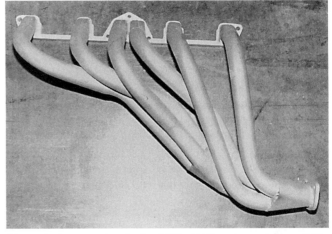

Nissan factory race header for 1970—78 Z-car was once available from Nissan Motorsports. Stainless header is for use with square exhaust ports. Header is no longer available (NLA).

Supplier List

CAMSHAFTS & VALVE TRAIN

Crane Camshaft
P.O. Box 160-2
Hallandale, FL 33009
305/927-4261

Crower Cams and Equipment Co.
3333 Main Street
Chula Vista, CA 92011
619/422-1191

Iskenderian Racing Cams
16020 South Broadway
Gardena, CA 90248
310/770-0930

Nissan Motorsports
P.O. Box 191
Gardena, CA 90247
310/538-2610

CARBURETION

Mikuni American Corp.
8910 Mikuni Ave.
Northridge, CA 91324
818/885-1242

Dellortto Carburetors
Rivera Engineering
7928 South Greenleaf
Whittier, CA 90602
213/693-4273

Nissan Motorsports
P.O. Box 191
Gardena, CA 90247
310/538-2610

Weber Carburetors
Redline, Inc.
19701 Magellan Drive
Torrance, CA 90509
310/538-3232

CONNECTING RODS

Carrillo Industries
990 Calle Amanecer
San Clemente, CA 92613
949/498-1800

Crower Cams and Equipment Co.
3333 Main Street
Chula Vista, CA 92011
619/422-1191

Nissan Motorsports
P.O. Box 191
Gardena, CA 90247
310/538-2610

CRANKSHAFTS

Moldex Crankshaft Co.
25249 West Warren Avenue
Dearborn Heights, MI 48127
313/561-7676

Nissan Motorsports
P.O. Box 191
Gardena, CA 90247
310/538-2610

Velasco Crankshaft Service
12422 Benidict Avenue
Downey, CA 90242
213/862-3110

CYLINDER HEADS

B.C. Geralomy
3250 Monier Circle
Rancho Cordova, CA 95670
916/638-9008

Cylinder Head Abrasives
3250 Monier Circle
Rancho Cordova, CA 95670
800/456-5474

Nissan Motorsports
P.O. Box 191
Gardena, CA 90247
310/538-2610

Don Preston Racing
331 Pattie
Wichita, KS 67211
316/263-4769

Slovers's Porting Service
13231 Sherman Way 7
North Hollywood, CA 91605
818/764-2260

FLYWHEELS & CLUTCHES

Design Products
5462 Oceanus Drive, Unit E
Huntington Beach, CA 92649
714/892-1513

HKS/USA, Inc.
2801 E. 208th Street
Carson, CA 90810
310/763-9600

Nissan Motorsports
P.O. Box 191
Gardena, CA 90247
310/538-2610

Quarter Master Industries, Inc.
185 Lively Blvd.
Elk Grove Village, IL 60007
312/593-8999

Tilton Engineering
P.O. Box 1787
Buellton, CA 93427
805/688-2353

OILING SYSTEMS

Aviad Metal Products
7570 Woodman Pl. A-3
Van Nuys, CA 91405
213/786-4025

Armstrong Race Engineering (ARE)
8848 Steven Avenue
Orangevale, CA 95662
916/987-7629

Design Products
5462 Oceanus Drive, Unit E
Huntington Beach, CA 92649
714/892-1513

Nissan Motorsports
P.O. Box 191
Gardena, CA 90247
310/538-2610

Peterson Fluid Systems
7200 E. 54th Place
Commerce City, CO 80005
800/926-7867

Weaver Brothers
14258 Aetna Street
Van Nuys, CA 91041
818/785-3622

PISTONS

Arias Pistons
13420 South Normandie Avenue
Gardena, CA 90247
310/770-0055

Cosworth Pistons
C/O Malvern Racing
P.O. Box 33
Ivy, VA 22945
804/971-9688

Forged True Pistons
14131 East Freeway Drive
Santa Fe Springs, CA 90670
562/921-8418

JE Pistons
15312 Connector Lane
Huntington Beach, CA 92649
714/898-9763

Jahns Pistons
1360 N. Jefferson
Anaheim, CA 92807
714/579-3795

Nissan Motorsports
P.O. Box 191
Gardena, CA 90247
310/538-2610

Ross Racing Pistons
625 S. Douglas
El Segundo, CA 90245
310/536-0100

Venolia Pistons
2160 Cherry Industrial Circle
Long Beach, CA 90805
562/636-9329

PISTON RINGS

Arias Pistons
13420 South Normandie Avenue
Gardena, CA 90247
310/770-0055

Cosworth Pistons
C/O Malvern Racing
P.O. Box 33
Ivy, VA 22945
804/971-9688

Deves
Janasz Corporation
137-139 Nevada Street
El Segundo, CA 90245
310/322-8483

Nissan Motorsports
P.O. Box 191
Gardena, CA 90247
310/538-2610

Total Seal Piston Rings
2225 West Mountain View, Suite 16
Phoenix, AZ 85021
602/242-9421

Venolia Pistons
2160 Cherry Industrial Circle
Long Beach, CA 90805
562/636-9329

PLUMBING

Earl's Supply
14611 Hawthorne Blvd.
Lawndale, CA 90260
310/772-3605

Aeroquip Corp. Industrial Division
300 South East Avenue
Jackson, MI 49203-1972
517/787-8121
Note: Contact offices for name
of local distributor

  Chicago Office
  7320 West North Avenue
  Chicago, IL 60635
  312/453-6730 or 453-6731

  Hartford Office
  10 North Main Street
  West Hartford, CT 06107
  203/232-1447

Goodridge USA
20309 Gramercy
Torrance, CA 90501
310/533-1924

Xtreme Racing Products
5630 Imperial Highway
South Gate, CA 90280
562/861-4765

Russel Performance Products
20420 South Susana Road
Carson, CA 90745
310/639-7151

RACING ENGINES, COMPLETE

Don Preston Racing
331 Pattie
Wichita, KN 67211
316/263-4769

Robello Racing Engines
110 - 2nd Avenue South, Unit D8
Pacheco, CA 94553
925/682-0103

Sunbelt Performance Engines
730 N. Price Rd.
Sugar Hill, GA 30518
770/932-0160

Parts List

CYLINDER BLOCK
The 6-cylinder L-series cylinder block, introduced in 1969, went through a number of evolutionary changes during its 14 year (US market) production run. The L26 block (8/73-11/74) retains the same 83mm/3.268" bore as the L24 but is notched for the larger (38mm/1.38") exhaust valve. The L28 block (from 12/74) has an increased bore (86mm/3.386"), siamesed cylinders, a slightly relocated camshaft oiling hole and a strengthened crankcase area. Centerline of the crankshaft to the head gasket surface is 207.85mm/8.18" (±0.05mm/0.002"). Total cylinder block height (gasket surface to gasket surface) is 265.85mm/10.467" (±0.15mm/0.005"). New cylinder blocks are NLA from NMC.

H/D MAIN CAP BOLT
These heavy-duty bolts are constructed from a higher grade material which allows a 50% increase (to 60 ft/lb) in the torque setting. Heavy-duty bolts are marked with a small 12'.

12293-VO700	H/D Main Bearing Cap Bolt	14

CRANKSHAFT
All 6-cylinder L-series crankshafts are OE units featuring 8 counterweights. Stroke dimension is given in ( ). Early L24 engines (up to L24-003606) have a 6 counterweight crankshaft and should be upgraded to an 8 counterweight unit prior to any high-performance modifications. The roller type pilot bearing replaces the OE solid bronze unit.

12201-E3002SV	L20A Crank (69.7mm/2.74")	1
12201-E3101SV*	L24 Crank (73.7mm/2.90")	1
12201-P3000SV*	L26/L28 Crank (79mm/3.11")	1
12201-VO790*	LD28 Crank (83mm/3.27")	1
32202-09500	Roller Pilot Bearing	1

OPT DRIVE GEAR & SPINDLE
Deterioration of the surface hardness of the OE steel drive gear can lead to premature wear when running at sustained high RPM. The improved wear characteristics of a bronze drive gear make this a must for all high-performance engines. New bronze drive gear and spindle must be installed as a set.

15043-73400	FL Drive Gear	1
15040-21001	OE Spindle	1

CONNECTING ROD
All L-series connecting rods are OE units featuring 9mm rod bolts. The center-to-center length is given in ( ). Early L24 engines (up to L24-096180) have 8mm rod bolts. Later style connecting rods with 9mm bolts should be installed prior to any high-performance modifications. Connecting rod bolts should be replaced during each rebuild.

12100-84GOO*	L24 Con Rod (133mm/5.24")	6
12100-16G10*	L26/L28 Con Rod (130.5mm/5.13")	6
12109-NO100	9mm Connecting Rod Bolt	12
12112-V5201	9mm Connecting Rod Nut	12

ROD & MAIN BEARINGS
These high-performance rod and main bearing shells utilize a tri-metal design featuring a copper-lead alloy sandwiched between a steel shell and an electro-plated surface layer (babbitt). Various thicknesses are available so that the proper oil clearance can be obtained when assembled with a standard size (uncut) L-series crankshaft. The additional oil clearance will result in increased lubrication, cooling, RPM and power. Dimension indicates bearing shell thickness.

12111-22010	.0585 " Rod Bearing	12
12111-22011	.0588" Rod Bearing	12
12111-22012	.0592 " Rod Bearing	12
12111-22013	.0580" Rod Bearing	12
12111-22014	.0578" Rod Bearing	12
12215-22010	.0714" Main Brng/Lower (#1/#7)	NLA
12215-22011	.0718" Main Brng/Lower (#1/#7)	NLA
12215-22012	.0722 " Main Brng/Lower (#1/#7)	NLA
12215-22013	.0710 " Main Brng/Lower (#1/#7)	2
12215-22014	.0707 " Main Brng/Lower (#1/#7)	2
12216-22010	.0714" Main Brng/Upper (#1/#7)	4
12216-22011	.0718 Main Brng/Upper (#1/#7)	4
12216-22012	.0722 " Main Brng/Upper (#1/#7)	4
12216-22013	.0710 " Main Brng/Upper (#1/#7)	2
12216-22014	.0707 " Main Brng/Upper (#1/#7)	2
12231-22010	.0714" Main Brng (#2/#3/#5/#6)	8
12231-22011	.0718" Main Brng (#2/#3/#5/#6)	8
12231-22012	.0722 " Main Brng (#2/#3/#5/#6)	8
12231-22013	.0710 " Main Brng (#2/#3/#5/#6)	8
12231-22014	.0707 " Main Brng (#2/#3/#5/#6)	8
12247-22010	.0714 " Main Brng (center)	2
12247-22011	.0718" Main Brng (center)	2
12247-22012	.0722 " Main Brng (center)	2
12247-22013	.0710 " Main Brng (center)	2
12247-22014	.0707 " Main Brng (center)	2

CRANKSHAFT
The VG30DE crankshaft has a stroke of 83mm/3.27".

12200-30P20*	VG30DE Crankshaft	1

CONNECTING ROD
The center-to-center length of the VG30DE connecting rod is 154.15mm/6.07". Connecting rod bolts should be replaced during each rebuild.

12100-30P00*	VG30DE Connecting Rod	6
12109-30P00*	Connecting Rod Bolt	12
12112-40F00*	Connecting Rod Nut	12

VG30DE BILLET SET
Cam lobes are unground.

13001-RR829	VG30DE Billet Set	1

OPTIONAL TURBO INJECTOR
When increasing boost, or adding a nitrous system, the OE injectors no longer provide a sufficient amount of fuel for the VG30DETT engine. These 555cc (vs 360cc) NISMO injectors will provide the additional fuel requirement.

16600-RR701	555cc Turbo Injector (to 9/94)	6
16600-RR541	555cc Turbo Injector (from 9/94)	6

K&N FILTERCHARGER KIT
Replaces OE air plenum/filter assy with a low-restriction K&N unit for improved throttle response and power throughout the entire RPM range. Kit includes K&N conical filter assy, a precision engineered aluminum adapter, all necessary mounting hardware and service kit. 50 state legal. Replacement service kit (99996-KNKIT) includes additional filter oil (8oz) and cleaner fluid (12oz) necessary for servicing the K&N filter.

99996-Z32KN	Filtercharger Kit	1
99996-KNKIT	Filter Oil & Cleaner	1

EURO TURBO RADIATOR
Due to high-speed driving conditions the European turbo radiator is 16% more efficient than the US version.

21410-37POO	Turbo Radiator (M/T)	1
21460-37POO	Turbo Radiator (AT)	1

EURO TURBO OIL COOLER KIT
Designed for the European 300ZX turbo, this larger (175mm x 325mm x 50mm) oil cooler will dissipate almost 300% more heat than the OE (98mm x 285mm x 32mm) oil cooler. Kit includes cooler, mounting bracket and required oil tube and connecting hoses.

21305-37P00K	Oil Cooler Kit	1

BORLA EXHAUST SYSTEM
Increased power, great sound, aggressive styling, 50 state legal, aircraft quality stainless steel construction and a perfect fit from the cat back. On top of all that it's almost 15 lb lighter than the OE system and takes less than one hour to install using the OE hangers.

99996-15320	Exhaust System	1
99996-14339	2+2 Exhaust System	1

CLUTCH ASSY
Due to high demand, Nissan Motorsports is once again offering the NISMO clutch cover (or pressure plate) and disc. The NISMO 240mm clutch cover is rated at 800kg (OE is 580kg). The Centerforce Dual Friction® clutch assembly includes both the clutch cover and disc. The Centerforce clutch cover uses a series of counterweights to increase (up to 90%) the clamping force as RPM's rise. All clutches require OE T/O bearing (30502-45P00).

240MM (MON-TURBO)
30210-RS560	NISMO Clutch Disc	1
30100-RN580	NISMO Disc	1
99996-240DF	Centerforce Clutch Assy	1

250MM (TURBO)
99996-250DF2	Centerforce Clutch Assy	1

SPEEDOMETER PINION
To compensate for tire, wheel or differential gearing changes, an alternate speedo pinion may have to be installed to correct the speedometer. Increasing the tooth count on the pinion will decrease the MPH readout

32743-30P17*	17t Speedo Pinion (M/T)	1
32743-30P18*	18t Speedo Pinion (M/T)	1
32743-30P19*	19t Speedo Pinion (M/T)	1
32743-30P20*	20t Speedo Pinion (M/T)	1
32743-30P21	21t Speedo Pinion (M/T)	N/A
32743-30P22*	22t Speedo Pinion (M/T)	1
32743-36F17	17t Speedo Pinion (A/T)	1
32743-36F18*	18t Speedo Pinion (A/T)	1
32743-36F19*	19t Speedo Pinion (A/T)	1
32743-36F20*	20t Speedo Pinion (A/T)	1
32743-36F21	21t Speedo Pinion (A/T)	1
32743-36F22	22t Speedo Pinion (A/T)	1
32743-35F17*	17t Speedo Pinion (Turbo A/T)	1
32743-35F18*	18t Speedo Pinion (Turbo A/T)	1
32743-35F19	19t Speedo Pinion (Turbo A/T)	1
32743-35F20*	20t Speedo Pinion (Turbo A/T)	1
32743-35F21	21t Speedo Pinion (Turbo A/T)	1
32743-35F22	22t Speedo Pinion (Turbo A/T)	1
32743-35F23	23t Speedo Pinion (Turbo A/T)	1

STABILIZER BARS
All of these OE light-weight stabilizer bars utilize existing mounting brackets and end link assemblies. Refer to the Nissan factory service manual before ordering.

54611-33P20*	27.2mm Bar / F	1
54613-33P00*	27.2mm Bushing / F	2
54611-33P21*	28.6mm Bar / F	1
54613-33P10*	28.6mm Bushing / F	2
56230-33P06*	15.9mm Bar / R	1
54613-16V00*	15.9mm Bushing / R	2
56230-33P20*	21.0mm Bar / R	1
54613-04F0l*	21.0mm Bushing / R	2
56230-33P10*	25.4mm Bar / R	1
54613-04F03	25.4mm Bushing / R	2

H/D END LINK BUSHINGS
Replacing the worn OE rubber bushings with long lasting urethane low-deflection bushings will significantly improve the effectiveness of the rear stabilizer bar. Dimensions are 1.125" (OD), .400" (ID) and .625" (T).

99996-EG800	Bushing	8

KONI SHOCK ABSORBERS
The KONI Sport shocks are adjustable, twin-tube, low-pressure gas units. The front shock is externally adjustable. Installation of KONI Sport shocks will de-activate the electronic suspension.

99996-1118S	KONI Sport Shock / F	2
99996-1119S	KONI Sport Shock / R	2

EIBACH SPRING SET
Eibach high-performance springs are a direct replacement and will lower the ride height approximately 3/4". The progressively wound non-turbo springs are rated at 171/314 lb (front) and 142/228 lb (rear). The turbo springs, also progressively wound, are rated at 185/250 lb (front) and 145/200 1 (rear).

99996-S2068	Spring Set	1
99996-63201	Spring Set (turbo)	1

OPT HID BUSHINGS
Optional NISMO bushings are manufactured from a material more rigid than OE to minimize suspension deflection.

54476-RS590	Tension Rod Bushing (#1)	2
54560-RS590	Lower Control Arm Bushing (#2)	2
55000-RS580	Bushing Set (includes #3 & #4)	1
55152-RS520	Bushing (#5/HICAS only)	2
55157-RS580	Bushing (#6/HICAS only)	2
56219-RS580	Bushing (#7)	2
56217-RS580	Shock Rod Bushing/Upper (#8)	2
56218-RS580	Shock Rod Bushing/Lower (#9)	2
55401-RS590	Bushing Set (includes #10 thru #13)	1

BRAKE PADS (HIGH-PERFORMANCE)
Repco Metal Master high-performance semi-metallic (nonasbestos) brake pads provide improved stopping power under all types of driving conditions. Other benefits include a rotor-friendly pad material, firmer brake pedal, no squeal and minimal dusting.

99996-D1170M	Repco Pad Set / F	1
99996-D461M	Repco Pad Set / R	1

BRAKE PADS (COMPETITION)
Hawk competition (blue) brake pads are an excellent raceproven design. Due to a high operating temperature (600' to 1400°), these are recommended as a RACE ONLY item.

41060-HB178	Hawk Pad Set / F	1
44060-HB179	Hawk Pad Set / R	1

SKYLINE GT BRAKE SYSTEM
The Skyline GT brake rotors are cross-drilled and unidirectional for increased cooling. Front rotors are 295mm (vs 280mm) in diameter and 32mm (vs 26/30mm) thick. Rear rotors maintain OE dimensions. Installation also requires listed longer brake pad retaining pins (115.7mm vs. 112mm) and alternate brake tubes. Cross-drilled brake rotors will increase pad wear.
CAUTION: Some grinding may be necessary to obtain proper wheel/caliper clearance.

40206-05U12	Rotor / RF	1
40206-05U13	Rotor / LF	1
41001-05U01	Caliper / RIF	1
41011-05U01	Caliper / LF	1
43206-05U12	Rotor / RR	1
43206-05U13	Rotor / LR	1

41217-40P00	Retaining Pin	4
46245-05U01	Alt Brake Tube / RF	1
46246-05U01	Alt Brake Tube / LF	1
41120-30P25	Caliper Rebuild Kit	1

H/D DECK LID STRUTS
A must when installing a rear wing or spoiler on the nonturbo 300ZX. Also, lifting action will not deteriorate in cold weather. Will not fit the 2+2 model.

90450-30P11	Strut Assy	2

7' FRONT PANEL EMBLEM
This attractive 'Z' emblem is OE on the non-USA 300zx. Easily installed on 300ZXs produced up to 7/90. 300ZXs produced after 7/90 will require the indentation to be filled in or optional center panel (62310-50P00) listed below.

62889-40P00	'Z' Emblem	1
62310-50P00	Opt Center Panel	1

'NISSAN FAIRLADY Z' EMBLEM
The Japan market Nissan sports car (since 1961) has been known as the 'Fairlady'. This attractive OE emblem replaces the 'Nissan' on the left rear hatch/trunk.

93094-30P01	'Nissan Fairlady Z' Emblem	1

HELLA AIR HORNS
Powerful and penetrating enough to get attention over any level of road or race track noise. Kit consists of 2 trumpets, high-performance compressor, relay and all mounting hardware.

99996-85105	Hella Air Horn Kit	1

CRANKSHAFT
The VG30E crankshaft has a stroke of 83mm/3.27". The later crankshaft (12201-21V80) has a 18.7mm/.736" longer snout and requires the later type pulley (including bolt and washer).

12201-02P81*	VG30E Crankshaft (to 4/87)	1
12200-21 V11*	VG30E Crankshaft (from 4/87)	1

CONNECTING ROD
The center-to-center length of the VG30E connecting rod is 54.15mm/6.07". The later W connecting rod (121001VO2) requires the W piston with full floating pin. Connecting rod bolts should be replaced during each rebuild.

12100-02P01	VG30E Con Rod (to 4/87)	6
12100-0W002*	VG30E Con Rod (from 4/87)	6
12109-V5211	Connecting Rod Bolt	12
12112-V5201	Connecting Rod Nut	12

OPT RACE CYLINDER HEADS
These especially designed cylinder heads have relocated intake and exhaust ports for maximum air flow and smaller combustion chambers for increased compression. Requires fabricated intake and exhaust systems and some dedicated valve train components. Legal for IMSA GT and some offroad racing classes.

11041-V52RN	Race Cylinder Head / RH	1
11091-V52RN	Race Cylinder Head / LH	1
99996-V5001	Lifter Cup	12
99996-V5002A	Pushrod	12
99996-V5003	Rocker Arm	12
99996-V52HF	Header Flange	2

VG30E CAMSHAFTS
The Euro turbo cams have same lift (.392") as the OE cams, however, the overall duration is increased resulting in a 10HP increase in the VG30ET engine (slightly less in the nonturbo). Basic cam specs are indicated in ( ). All new cam installations require new OE valve lifters (13231-V5014).

13001-07P80	Euro Cam / RH (.392"/264°)	1
13061-07P80	Euro Cam / LH (.392"/264°)	1
99996-V256H	Isky Cam Set #1 (.425"/256°)	1
99996-V262H	Isky Cam Set #2 (.430"/264°)	1

VG30E BILLET SET
Cam lobes are unground.

99996-RN853M	VG30E Billet Set	1

HEAD GASKET
NISMO 89mm metal head gasket set. Compressed thickness is 1 mm.

11044-RR810	Head Gasket Set	1

OPT AIR INTAKE SYSTEM
These parts will increase the volume of air through the primary portion of the intake system. The intake screen (OE on turbo models from 9/86) replaces the solid intake duct behind the RH headlamp assy. The intake duct replaces the restrictive OE unit which is located directly under the upper radiator hose.

62536-07P00	Air Intake Screen	1
62860-01P01	Air intake Duct	1

K&N FILTERCHARGER KIT
Replaces OE air plenum/filter assy with a low-restriction K&N unit for improved throttle response and power throughout the entire RPM range. Kit includes K&N conical filter assy, all necessary mounting hardware and service kit. 50 state legal. Replacement service kit (99996-KNKIT) includes additional filter oil (8oz) and cleaner fluid (12oz) necessary for servicing the K&N filter.

99996-Z31KN	Filtercharger Kit	1
99996-KNKIT	Filter Oil & Cleaner	1

H/D RADIATOR
Due to high-speed driving conditions the European turbo radiator is 10% more efficient than the US version. Fits all models.

21450-07P02	H/D Radiator	1

OIL PUMP
The VG30E oil pump is an inner gear type which is driven directly off the crankshaft. The turbo pump increases the volume of oil 25% over the non-turbo pump.

15010-07P02*	Turbo Oil Pump (to 3/87)	1
15010-07P03*	Turbo Oil Pump (from 3/87)	1

RACE OIL PUMP GEAR SET
This chromoly steel replacement oil pump gear set is recommended for use when continuous engines speeds exceed 7500RPM.

99996-V1010	Oil Pump Gear Set	1

ADAPTER & REMOTE BLOCK
The adapter (15238-GCO01) replaces the existing oil filter and requires the installation of the remote oil filter mount (99996-T101X) when installing an aftermarket oil cooler. The remote oil filter mount has provisions for an oil temperature sending unit. Requires use of Fram HP4 or equivalent oil filter. Both the adapter and remote oil filter mount have 1/2 NPT female threads.

15238-GC001	Adapter	1
99996-T101X	Remote Mounting Block	1

BORLA EXHAUST SYSTEM
Increased power, great sound, aggressive styling, 50 state legal, aircraft quality stainless steel construction and a perfect fit. Replaces rear muffler assembly. Non-turbo models require the larger turbo cat exit pipe.

99996-12120	Exhaust System	1
20030-02P00*	Cat Exit Pipe (to 9/85)	1
20030-19P05*	Cat Exit Pipe (from 9/85)	1

CLUTCH ASSY
Due to high demand, Nissan Motorsports is once again offering the NISMO clutch cover (or pressure plate) and disc. Both the NISMO 240mm and 250mm clutch covers are rated at 800kg. OE specs are 580kg (non-turbo to 9/86), 500kg (non-turbo from 9/86) and

600kg (turbo). The race clutch disc (9/6-E3045) is a solid-hub bronze-button unit designed for competition use only. The Centerforce Dual Friction® clutch assembly includes both the clutch cover, and disc. The Centerforce clutch cover uses a series of counterweights to increase (up to 90%) the clamping force as RPM rises. The Tilton aluminum flywheel (with steel insert) weighs 11lb. All clutches require OE T/O bearing (3050 1000).

240MM (ALL NON-TURBO & TURBO TO 9/86)
30210-RS560	NISMO Clutch Cover	1
30100-RN580	NISMO Disc	1
99996-E3045	Race Disc	1
99996-240DF	Centerforce Clutch Assy	1
9999641012	Tilton Flywheel	1

250MM (TURBO FROM 9/86)
3021 0-RS910	NISMO Clutch Cover	1
30100-22P60*	OE Disc	1
99996-250DF1	Centerforce Clutch Assy	1

R200 RING & PINION SETS (10MM)
The ring gear has an OD of 200mm (R200), an ID of 128mm and uses a 10mm ring gear bolt with a 1.25 thread pitch.

38100-P6200*	3.364 R&P Set	1
38100-P6300*	3.545 R&P Set	1
38100-P6400	3.700 R&P Set	1
38100-P6500	3.900 R&P Set	1
38100-P6600	4.111 R&P Set	1

R200 RING & PINION SETS (12MM)
The ring gear has an OD of 200mm (11200), an ID of 128mm and uses a 12mm ring gear bolt with a 1.25 thread pitch. The special 12mm bolt are 22mm in length and drilled for safety wire.

38100-12S01	3.364 R&P Set	1
38100-13S01*	3.545 R&P Set	1
38100-14S01*	3.700 R&P Set	1
38100-15S00*	3.900 R&P Set	1
38100-16S01*	4.111 R&P Set	1
38102-RN800	Ring Gear Bolt (1.25mm)	10

R200 LIMITED-SLIP ASSY
The R200 US assy is a 4-pinion unit with approximately 45 ft/lbs breakaway torque. Side gears are 30mm x 29t. All parts listed below the actual US assemblies are servicing components only.

38420-RS650	US Assy (10mm)	1
38420-RS660	US Assy (12mm)	1
38423-RR650	Side Gear	N/A
38225-N3100*	Circlip	2
38425-C6000*	Pinion Gear	4
38427-N3210	Pinion Shaft	1
38431-N3210	Pressure Ring	2
38432-N3210	Friction Plate (1.75mm)	4
38433-N3210	Friction Disc (1.75mm)	4
38433-N3211	Friction Disc (1.85mm)	(4)
38434-N3210	Screw	4
38435-N3210	Spring Plate (1.75mm)	2
38436-N3210	Spring Disc (1.75mm)	2

SPEEDOMETER PINION
To compensate for tire, wheel or differential gearing changes, an alternate speedo pinion may have to be installed to correct the speedometer. Increasing the tooth count on the pinion will decrease the MPH readout. P/N includes speedo pinion and sleeve assy.

32702-58S16*	16t Speedo Pinion (M/T)	1
32702-58S17*	17t Speedo Pinion (M/T)	1
32702-58S18*	18t Speedo Pinion (M/T)	1
32702-58S19*	19t Speedo Pinion (M/T)	1
32702-58S20*	20rt Speedo Pinion (M/T)	1
32702-E9821	21t Speedo Pinion (M/T)	1
32702-58S22*	22t Speedo Pinion (M/T)	1
32702-E9823	23t Speedo Pinion (M/T)	1
32703-V1017*	17t Speedo Pinion (A/T)	1
32703-V1018*	18t Speedo Pinion (A/T)	1
32703-V1019*	19t Speedo Pinion (A/T)	1
32703-V1020*	20t Speedo Pinion (A/T)	1
32703-VI 021*	21t Speedo Pinion (A/T)	1
32703-V1022*	22t Speedo Pinion (A/T)	1
32703-V1023	23t Speedo Pinion (A/T)	1

DIFFERENTIAL OIL COOLER
For any type of sustained high-speed driving a differential oil cooler should be installed to reduce the heat buildup and prolong the life of the ring & pinion, US assy, bearings and seals. The pump has a maximum flow rate of 1.6 GPM. Full plumbing should include a -10 line into and a -8 line out of the pump and a 75 micron filter mounted between the differential and the pump assy.

21660-37P00	Pump Assy	1
38501-E8775	Oil Cooler	1
38351-P9010	Opt Rear Cover	1

STABILIZER BARS (FRONT)
Except where noted, all factory front stabilizer bars utilize existing mounting brackets and end link assemblies. The 1 1/8" stabilizer bar kit by Suspension Techniques, however, includes urethane bushings and all mounting hardware.

5461 1-01P10*	22mm Bar	1
54613-01 P20*	22mm Bushing	2
54611-01P00*	23mm Bar	1
54613-W1410*	23mm Bushing	2
54611-07P00	24mm Bar	1
54613-07P00	24mm Bushing	2
54611-23P01	26mm Bar	1
54613-23P01	26mm Bushing	2
54614-23P00	26mm Special Bracket	2
99996-R4302	1 1/8" Stabilizer Bar Kit	1

STABILIZER BARS (REAR)
All factory rear stabilizer bars utilize existing mounting brackets and end link assemblies. The 1" stabilizer bar kit by Suspension Techniques, however, includes urethane bushings and all mounting hardware.

56230-01P10*	22.2mm Bar	1
54613-04F01*	22.2mm Bushing	2
56230-04F12*	24mm Bar	1
54613-04F02*	24mm Bushing	2
56230-04F13	25mm Bar	1
54613-04F03	25mm Bushing	2
99996-R4352	1" Adj Stab Bar Kit	1

H/D END LINK BUSHINGS
Replacing the worn OE rubber bushings with long lasting urethane low-deflection bushings will significantly improve the effectiveness of the stabilizer bars. The front and rear bars require eight (8) bushings each. However, from 9/86, the front bar requires only four (4) bushings. Dimensions are 1.125" (OD), .400 (ID) and .625 (T).

99996-EG800	Bushing	12/16

KOMI SHOCK ABSORBERS
The KONI Sport insert is an (externally) adjustable, twintube, low-pressure gas unit while the Sport shock is an adjustable monotube, high-pressure gas unit. Installation of KONI Sport inserts and shock absorbers will cle-activate the electronic suspension.

99996-1143S	KONI Sport Insert / F	2
99996-1248S	KONI Sport Shock / R	2

STRUT BUMP STOP
Keeps the strut insert from bottoming out on a lowered 300ZX. Can be trimmed to fit any application. Sold by the pair.

99996-6104G	Bump Stop	1

S/S SPRINGS (STAGE I)
These OE springs from the limited production 300ZX Turbo S/S are a direct replacement and will significantly reduce body roll while maintaining full suspension travel. The

springs are rated at 207 lb/in (front) and 224 lb in (rear).

54010-26P00	S/S Spring / F	2
55020-26P00	S/S Spring / R	2

EIBACH SPRING SET (STAGE II)
Eibach high-performance springs are a direct replacement and will lower the ride height approximately 3/4". The progressively wound springs are rated at 142/240 lb in (front) and 183/257 lb/in (rear).

99996-63011	Spring Set	1

WHEEL STUDS
Longer studs will allow the installation of spacers and/or certain aftermarket alloy wheels. Early non-turbo models (to 9/86) have a 4-bolt pattern. Later non-turbo (from 9/86) and all turbo models have a 5-bolt pattern.

40222-RS025	Wheel Stud	16/20
40222-RS015	Wheel Stud	16/20

PART NO.	A	B	C	D	E	SIZE	SERRATION
40222-RS025	59	25	7	19	5	M12 x 1.25	13 x 36T
40222-RS015	50	14	7	19	5	M12 x 1.25	13 x 36T

BRAKE PADS (HIGH-PERFORMANCE)
Repco Metal Master high-performance semi-metallic (nonasbestos) brake pads provide improved stopping power under all types of driving conditions. Other benefits include a rotor-friendly pad material, firmer brake pedal, no squeal and minimal dusting.

99996-R7010M	Repco Pad Set / F	1
99996-D1152M	Repco Pad Set / F (turbo from 9/86)	1
99996-N7020M	Repco Pad Set / R	1

AIR DAM
Kaminari front air dams are gel-coated for easy painting and include instructions and mounting hardware.

99996-R6002	Air Dam ('84 & '85)	1
99996-R6003	Air Dam ('86 & 50th)	1

REAR WING
This fiberglass rear wing is identical to the one used on SCCA and IMSA winning 300ZXs. To counteract front end lift, an air dam should also be installed. No mounting hardware. Also, will NOT fit 2+2s and requires modification to clear rear wiper.

99996-R8001	Rear Wing	1

H/D DECK LID STRUTS
A must when installing a rear wing, spoiler or shade kit on the 300ZX. Will not fit 2+2.

90450-01P10	Strut Assy	2

HELLA EURO HEADLAMPS
Hella Euro-spec lights provide a brighter light for a greater distance (over 5,000' on high-beam) than standard sealed beam units. It is suggested that you upgrade the light fuse by 5A or install a relay when installing replacement Hella headlamps. Hella 200mm rectangular Euro headlamps are equipped with 60/55W bulbs. Optional 100/80W H4 bulbs can be purchased separately. For the 300ZX/Z31 up to 9/86 only. OFF-ROAD USE ONLY!

99996-72206	Hella Headlamp	2
99996-78159	Opt 100/80W H4 Bulb	2

HELLA AIR HORNS
Powerful and penetrating enough to get attention over any level of road or race track noise. Kit consists of 2 trumpets, high-performance compressor and all mounting hardware.

99996-85105	Hella Air Horn Kit	1

RADIO PLATE
Blanking plate fits in dash when radio is removed for competition purposes.

28321-07P00	Radio Plate	1

HIGH-PERFORMANCE PISTONS
Compression increases in L-series engines should be obtained with the use of alternate pistons and NOT by cutting the cylinder head (OE thickness = 108mm/4.25"). The octane rating of currently available pump gasoline will limit the compression ratio to below 10:1. Chances are that your existing block and/or cylinder head have had material removed at some point in time. It is very important that the compression ratio be verified prior to final assembly (see 'Auto Math Manual' in literature section).

12010-H2711*	L24/L26 Piston (83mm/3.27")	6
12033-A8620*	L24/L26 Ring Set (83mm/3.27")	1
12010-H2714*	L24/L26 Piston (83.5mm/3.29")	6
12036-A8620*	L24/L26 Ring Set (83.5mm/3.29")	1
12010-H2716*	L24/L26 Piston (84mm/3.31")	6
12038-A8620*	L24/L26 Ring Set (84mm/3.31")	1
12010-E3118	L24/L26 Piston (84.5mm/3.33")	6
12040-E3100	L24/L26 Ring Set (84.5mm/3.33")	1
12010-P8612	L28 Piston (86mm/3.39")	6
12033-P8610	L28 Ring Set (86mm/3.39")	1
12010-P8615	L28 Piston (86.5mm/3.41")	6
12036-P8610	L28 Ring Set (86.5mm/3.41")	1
12010-P8617	L28 Piston (87mm/3.43")	6
12038-P8610	L28 Ring Set (87mm/3.43")	1

FORGED RACING PISTONS
It will be necessary for the piston dome and combustion chamber to be modified to achieve the desired (and verified) compression ratio. All forged racing pistons are designed to accept Deves rings (1/16" compression & 5/32" oil control) and require the use of the later L24 (12100-84G00) connecting rod. Each piston is packaged with an .826 diameter wrist pin and (2) Teflon buttons. When ordering individual pistons be sure to specify correct cylinder (#1/#4/#5 or #2/#3/#6). RACE APPLICATION ONLY!

99996-D1061B	L24/.040 Piston (84mm/3.307")	6
99996-M 1021	L24/.040 Ring Set (84mm/3.307")	1
99996-M1013B	L26/.040 Piston (84mm/3.307")	6
99996-M1021	L26/.040 Ring Set (84mm/3.307")	1
99996-N1010B	L28/.040 Piston (87mm/3.425")	6
99996-N 1021	L28/.040 Ring Set (87mm/3.425")	1

2.8 BIG BORE KIT
Increases displacement to 3098cc (193ci). Kit includes LD28 crankshaft, L24 connecting rods, 89mm pistons and rings. Pistons have a positive deck height of approximately .025 and must be machined to accommodate various block and/or cylinder head combinations. Compression ratio can be adjusted by using either of the two 91mm cylinder head gaskets.

99996-28BBK	2.8 Big Bore Kit	1
11044-91MM1	91 mm Head Gasket (T=1mm)	1
11044-91MM2	91mm Head Gasket (T=2mm)	1

CAMSHAFTS
New OE rocker arms (13257-W0300) are required with all new L-series cam installations. In addition, some cams may require additional components (❶, ❷ or ❸) to complete the installation. Lash pad listings are approximate and may vary. Basic cam specs are indicated in ( ). Race cams (L4 & L3) must use OE spray bar assy (13100-E3004*).

13001-N3626	Hi-Perf Cam (.434"/256°)	1
99996-E1036	L7 Cam (.475"/270°) ❶	1
99996-E1032	L9 Cam (.490"/290°) ❷	2
99996-E1054	L4 Race Cam (.580"/278°) ❸	3
99996-E1055	L3 Race Cam (.620"/300°) ❸	1

13257-W0300 OE Rocker Arm (ratio=1.5:1) 12
❶ 9/6-M1160
❷ 9/6-M1160, 9/6-M1046 & 9/6-M1152/M1050
❸ 9/6-M1280, 9/6-M1046 & 9/6-M1150/M1051

CAM BILLET
Cam lobes are unground. The undrilled cam billet is applicable to all production Z engines (up to L28-155909) and competition engines which use the external spray bar assy (13100-E3004). On later Z engines (from L28-155910) an internal oiling system similar to the 4-cylinder L-series engine was utilized.

99996-E1029 Cam Billet (undrilled) 1
13001-E4651 Cam Billet (drilled) 1

COMPETITION VALVE SPRING
Valve spring is rated at 280 lb/in, has an OD of 1.34" and coil binds at .970". Assembled height is 1.720" with a seat pressure of 120 1. Spring life is greatly reduced when cam lift exceeds .600 Requires use of optional aluminum or steel retainers.

99996-M1046 Valve Spring 12

VALVE SPRING RETAINERS
These special retainers are required when installing the competition valve spring (9/6-M1046). The lash pad thickness dictates which retainer (short, medium or tall) is utilized. The acceptable range of lash pad thickness is indicated in ( ).

99996-M1152 Aluminum Retainer (.150-.180) 12
99996-M1151 Aluminum Retainer (.190-.240) 12
99996-M1150 Aluminum Retainer (.250-.340) 12

99996-M1050 Steel Retainer (.150-.180) 12
99996-M1052 Steel Retainer (.190-.240) 12
99996-M1051 Steel Retainer (.250-.340) 12

VALVE LASH PADS
Due to a modified base circle, alternate lash pads are required when installing a reground camshaft. Always consult the cam grinder for the recommended lash pad required to correct the rocker arm geometry.

99996-MI 155 .150" Lash Pad 12
99996-M1160 .160" Lash Pad 12
99996-M1170 .170" Lash Pad 12
99996-M1180 .180" Lash Pad 12
99996-M1190 .190" Lash Pad 12
99996-M1200 .200" Lash Pad 12
99996-M1210 .210" Lash Pad 12
99996-M1220 .220" Lash Pad 12
99996-M1230 .230" Lash Pad 12
99996-M1240 .240" Lash Pad 12
99996-M1250 .250" Lash Pad 12
99996-M1260 .260" Lash Pad 12
99996-M1270 .270" Lash Pad 12
99996-M1280 .280" Lash Pad 12
99996-M1290 .290" Lash Pad 12
99996-M1300 .300" Lash Pad 12
99996-M1310 .310" Lash Pad 12
99996-M1320 .320" Lash Pad 12
99996-M1330 .330" Lash Pad 12
99996-M1340 .340" Lash Pad 12

ADJUSTABLE CAM SPROCKET
With 8 holes (OE is 3) this optional cam sprocket will provide a more precise setting of the cam timing. Adjustment is from -12* to +9' in 3' increments.

13024-E4621 Cam Sprocket 1

MARK	1	2	3	4	A	B	C	D
OE SPROCKET	0°	+4°	+8°					
13024-F4671	0°	+3°	+6°	+9°	-3°	-6°	-9°	-12°

H/D OE CAM SPROCKET
As a running production change in the early '70s, the early steel cam sprocket was replaced with a less expensive powdered metal piece. We have obtained quantity of these early (stronger) sprockets. The part to have when the rules won't allow the NISMO adjustable cam sprocket. Supply limited to quantity on hand.

13024-21000 H/D Cam Sprocket 1

H/D HEAD BOLTS
The turbo head bolts are constructed from a higher grade material which allows a torque spec of 65 ft/lb. A small '13' on the head of the bolt indentifies it as a turbo head bolt.

11056-P7600 Head Bolt (short) 9
11059-P7600 Head Bolt (long) 5
11058-21001 Head Bolt Washer 14

HEAD GASKET (L24/L26)
NISMO 85mm x 89mm high-performance composition type head gasket. Compressed thickness is 1.2mm.

11044-E4620 L24/L26 Head Gasket 1

HEAD GASKET (L28)
HKS 91mm steel sandwich type head gasket features a raised bead around each combustion chamber to ensure a tight seal. Dimensions in ( ) indicate compressed thickness.

11044-91MM1 Head Gasket (1mm) 1
11044-91MM2 Head Gasket (2mm) 1

HEAD GASKET (-O- RING)
NISMO composition type head gasket for use with 88.8mm (ID) sealing rings. Installation of sealing rings require machining of the cylinder block. Compressed thickness is 1.1mm.

11044 E4621 Head Gasket 1
11045-N3120 Sealing Ring 6

VALVES
The SI competition valve features stainless steel forgings, chrome stems, stellite tips and a swirl finish for improved flow characteristics. The optional valve seat (99996-N1121) is required when installing the larger 44mm/1.73" intake valve into the L24/L26 cylinder head. Valves will not work on L28 cylinder heads produced after 7/80.

13201-A1100* 42mm/1.65" Intake Valve 6
99996-M1030 SI 43mm/1.69" Intake Valve 6

13201-N4200* 44mm/1.73" Intake Valve 6
99996-N1100 SI 44mm/1.73" Intake Valve 6
99996-N1121 Opt 44mm/1.73" Valve Seat (6)

13202-N0401* 35mm/1.38" Exhaust Valve 6
99996-D1034 SI 35mm/1.38" Exhaust Valve 6

13207-H7210 Valve Seal 12

EURO BALANCE TUBE
Euro balance tube for 240Z (up to 6/72) has provisions for master-vac, cylinder block breather and vacuum control diaphragm (Arr) hoses only. Euro linkage shaft does not have arm for activating servo diaphragm.

14008-E4100 Balance Tube 1
16380-E4300 Linkage Shaft 1

44MM MIKUNI INDUCTION SYSTEM
With appropriate jetting and venturi sizing, this system can be applied to any 6-cylinder L-series engine. No air cleaner ssy is offered and some interference may exist between the rear carb and the brake master-vac unit. The connecting shaft between the firewall and intake manifold must be fabricated. Carb set includes 50mm stacks. Main 175 / Air 250 / Pilot 55 / Choke 37mm.

16010-E4620M 44mm Carb Set 1
14002-E4620 Intake Manifold 1
14035-E4610* OE Manifold Gasket 1
99996-E1044 Opt Header Gasket (1)
14330-E4620 Heat Shield 1
16174-20100 Insulator 6
16176-20101 Insulator Gasket 12
16360-E4675 Linkage Assy 1

17520-E4620	Fuel Tube	1
99996-S1044	Mikuni Carb Rebuild Kit	3

50MM MIKUNI INDUCTION SYSTEM

With appropriate jetting and venturi sizing, this induction system can be applied to any 6-cylinder L-series competition engine. SCCA GT-2 legal. No air cleaner is offered. The connecting shaft between the firewall and intake manifold must be fabricated. Main 180 / Air 170 / Pilot 70 / Choke 43mm. RACE APPLICATION ONLY!

16010-RR610	50mm Carb Set	1
14002-E4621	Intake Manifold	1
14035-E4621	Manifold Gasket	1
99996-E1044	Opt Header Gasket	(1)
14330-E4620	Heat Shield	1
16174-E4622	Insulator (Bakelite)	6
16176-14610	Insulator Gasket	12
16360-E4675	Linkage Assy	1
17520-E4620	Fuel Tube	1

COMPETITION FUEL PUMP

This 12V (negative ground) electric fuel pump has an unrestricted fuel pressure of 6.5 to 7.5 PSI and requires an aftermarket regulator. Maximum flow rate is 45GPH. Pump includes (2) mounting insulators, ground wire, special 400 micron filter and 1/4-18 female inlet/outlet. Blanking plate is required when removing OE fuel pump.

17010-A7600A	Electric Fuel Pump	1
16404-479808	Replacement 400 Micron Filter	1
17033-E4200	Mounting Bracket (to 6/72)	1
16420-E3020	Blanking Plate	1

EURO DAMPER

A single groove crank damper which reduces the rotational mass of the multi-groove OE unit. Damper is degreed to +20'. Does not allow the use of the air pump or A/C. Legal for SCCA's ITS class as OA diameter remains unchanged. Should be installed with special bolt/washer set (99996E1065) when used in competition. Total alignment difference between pulleys should not exceed 1/8".

12303-E4100	Euro Damper	1

RACE DAMPER

Combining the original concept with the latest production techniques, the classic L-series competition crankshaft damper is once again available. Featuring alloy steel construction and aerospace quality bonding with a new high heat and oil resistant elastomer, this fully degreed and balanced damper is finished in a durable black oxide. Can be used with OE W/P and alternator pulleys. However, it will not allow the use of the smog pump, A/C and/or P/S. Requires use of the special bolt/washer set (99996-E1065). Total alignment difference between pulleys should not exceed 1/8 ".

99996-E1060E	Race Damper	1

NISMO PULLEY SET

Crank damper is degreed to +45°. Must be installed as a set due to the narrow belt groove design. Does not allow the use of the smog pump, A/C and/or P/S. Requires use of the special bolt/washer set (99996-E1065). Total alignment difference between pulleys should not exceed 1/8".

12303-E4620	Crank Damper (102mm/4.1")	1
21051-E4620	W/P Pulley (102mm/4.1")	1
23150-E4621	Alternator Pulley (62mm/2.4")	1
99996-E8010	Special Belt	1

SPECIAL BOLT/WASHER SET

Set consists of special bolt which has been roll-threaded after the heat-treating process. The larger diameter washer has also been heat-treated. Together they will maintain the required torque (105 ft/lb) setting and prevent the crank damper from welding itself to the crankshaft. A must when building any high-performance or racing engine.

99996-E1065	Bolt/Washer Set	1

Z RACE HEADERS

Both headers are built with 18 gauge chrome plated steel and feature a 6-into-2 design with 29" primaries and 2.5" collectors. Primary tube size is indicated in ( ). Requires use of special manifold gasket (99996-E1044) listed below. Exhaust system from the collectors back must be fabricated.

99996-E 1141	Z Race Header #1 (1.625")	1
99996-E 1142	Z Race Header #2 (1.750")	1

MANIFOLD GASKET

These gaskets are designed only for cylinder heads with 11 square" exhaust ports. The L28 gasket is notched for injector clearance. Will not work with the OE exhaust manifold.

99996-E1044	L24/L26 Manifold Gasket	1
99996-N1044	L28 Manifold Gasket	1

OIL PUMP (HIGH-VOLUME)

Gear size is increased 5mm (40mm vs 35mm) which results in a 15% greater volume than non-turbo OE oil pump. Flow rate is 9.6 GPM at 3000 RPM. Pressure can be increased by installing either or both of the competition oil pump springs.

15010-S8000	High-Volume Oil Pump	1

OIL PUMP (COMPETITION)

Flow rate is 13.7 GPH at 4000 RPM. Over 90 PSI compensates for increased bearing clearance and/or the addition of an oil cooler. Includes optional competition inner and outer oil pump springs.

15010-A1110	Competition Oil Pump	1

COMPETITION OIL PUMP SPRINGS

Use of either or both of these springs will increase the pressure of the standard or turbo (high-volume) oil pump.

15133-E4620	Oil Pump Spring (inner)	1
15133-22010	Oil Pump Spring (outer)	1

OIL PUMP ADAPTER KIT

Allows pump to pull oil directly from the (modified) oil pan bypassing the small, restricted internal passages. Increases oil flow by 20%. Replaces the cover assy on the L-series oil pump and provides for the installation of an aftermarket oil cooler. Adapter kit includes the special pump cover, longer bolts and gasket. Requires the AN fitting set (-12 inlet & -10 outlet) to accept braided steel lines. Used with adapter (15238-GCO01) and remote oil filter block (99996-T101X) listed on page 19.

15015-GCO02	Oil Pump Adapter	1
99996-GCO03	AN Fittings	1

ADAPTER & REMOTE BLOCK

The adapter (15238-GCO01) replaces the existing oil filter and requires the installation of the remote oil filter mount (99996-T101X) when installing an after market oil cooler. The remote oil filter mount has provisions for an oil temperature sending unit. Requires Fram HP4 or equivalent oil filter. Both the adapter and remote oil filter mount have 1/2 NPT female threads.

15238-GC001	Adapter	1
99996-T101X	Remote Filter Mount	1

Z COMPETITION OIL PAN

This custom-made 8 qt oil pan is equipped with trap-door style baffling, windage tray and special oil pick-up. Provides constant oil supply under all racing conditions and reduces HP loss due to crankshaft and connecting rods rotating through the oil supply. Special gasket improves sealing between oil pan and cylinder block.

99996-E1131	Z Oil Pan	1
99996-E1135	Special Pan Gasket	1

ADAPTER
Replaces drain plug and allows the use of a Stewart-Warner mechanical oil temperature sending unit with any OE oil pan. Will not work with competition oil pan.

11128-1-12375	Adapter	1

FAN SHROUD & LOWER PANEL
This optional fan shroud improves efficiency of the cooling fan at low speeds. The lower panel helps keep unwanted ground-level turbulence from interfering with the air-flow through the radiator. Installing both parts together will enable the Z cooling system to operate at a more efficient level at any speed.

21475-E4102*	240Z Fan Shroud	1
74810-E4100*	Z Lower Panel (to 8/74)	1
74810-N4200*	Z Lower Panel (from 9/74)	1

EURO DISTRIBUTOR
This Hitachi (D606-52) single point distributor has a slightly quicker advance curve and was OE on the European 240Z. Requires use of the L24 distributor support (22178-21000) when installed on an L26 or L28 cylinder block. The distributor spring, cam and vacuum advance assy are service items only.

22100-E3101	Euro Distributor	1
22110-E3100	Dist Spring	1
22132-E3100	Dist Cam Assy (6°)	1
22132-E4600*	Opt Dist Cam Assy (12°)	(1)
22301-E3101	Vacuum Advance Assy	1
22178-21000*	L24 Distributor Support	1

SOLID DISTRIBUTOR PLATE
Eliminates vacuum advance mechanism of the OE (or Euro) single-point distributor and reduces high RPM point bounce. Includes OE point set.

99996-M2001	Dist Plate	1

COMPETITION WIRE SET
7mm solid core ignition wires that improve the spark conductivity of a competition prepared Z/ZX.
RACE APPLICATION ONLY!

99996-E2020	Wire Set	1

STARTER MOTOR
OE gear reduction starter is suitable for all high-performance applications.

23300-N5903R*	Gear Reduction Starter	1

CLUTCH ASSY
Due to high demand, Nissan Motorsports is once again offering the NISMO clutch cover (or pressure plate) and disc. The NISMO 225mm clutch cover is rated at 750kg and the NISMO 240mm clutch cover is rated at 800kg. Consult your Nissan service manual for OE specs. The race clutch disc (9/6-3040/45) is a solid-hub bronze button unit designed for competition use only. The Centerforce Dual Friction® clutch assembly includes both the clutch cover and disc. The centerforce clutch cover uses a series of co counterweights to increase (up to 90%) the clamping force as RPM rises. The Tilton aluminum flywheel (with steel insert) 11 lbs. All clutches require OE T/O bearing (30502-21000).

225MM (2-SEATER)

30210-RS600 †	NISMO Clutch Cover	1
30100-RN421	NISMO Disc	1
99996-E3040	Race Disc	1
99996-225DF †	Centerforce Clutch Assy	1
99996-F1011	Tilton Flywheel	1

† Requires later T/O bearing sleeve (30501-N1600) when installed on Z's produced up to 11/74.

240MM (2+2 & TURBO)

30210-RS560	NISMO Clutch Cover	1
30100-RN580	NISMO Disc	1
99996-E3045	Race Disc	1
99996-240DF	Centerforce Clutch Assy	1

ROLLER PILOT BEARING
This roller type pilot bearing replaces the OE solid bronze bushing.

32202-09500	Pilot Bearing	1

PRESSURE-PLATE COMPARISON

Application	ID/OD mm (in.)	Nissan Part Number	Installed Load kg (pounds)	Diaphragm Height mm (in.)
up to '73 510 (L16)	200 (7.87)/130 (2.12)	30210-23000	350 (744)	43-45 (1.69-1.77)
710 (L18)	"	30210-K0400	440 (882)	31-33 (1.522-1.30)
'73 & 74 620 (L16 & L18) & '73 610 (L18)	"	30210-23000	360 (794)	43-45 (1.69-1.77)
'75-77 710 (L20B)	"	30210-S0100	450 (992)	31-33 (1.22-1.30)
'75-76 610 (L20B)	"	"	"	"
'77-79 200SX (L20B)	"	30210-K0400	400 (882)	"
'78-81 510 (L20B & LZ20)	"	"	"	"
'80-83 200SX (LZ20 & LZ22)	"	"	"	"
'67-69 SRL311 (U20)	"	30210-20111	650 (1435)	43-45 (1.69-1.77)
'76-78 280ZX coupe & '79-83 non-Turbo coupe	225 (8.86)/150 (5.90)	30210-N4210	550 (1213)	33-35 (1.299-1.378)
'74-79 620 (L20B)	"	30210-P0100	"	"
'80-82 720	"	30210-Y0100	"	"
'73 240Z & '74 260Z	"	30210-N3100	"	43-45 (1.69-1.77)
'77 810	"	30210-Y0700	392 (888)	33-35 (1.30-1.38)
'78-80 810/'81-83 Maxima	"	30210-Y0600	"	"
Sports Option	"	30210-N3222	780 (1700)	"
'76-78 280Z 2+2	240 (9.45)/150 (5.90)	30210-P9201	"	"
'82 & 83 280ZX Turbo & 300ZX non Turbo	"	30210-P9510	550 (1213)	"
'84-85 300ZX Turbo	"	30210-P9600	600 (1323)	"
'80-85 720 4X4	"	30210-10W00	400 (882)	"

Metric Customary-Unit Equivalents

Multiply:	by:		to get:	Multiply:	by:		to get:

LINEAR
inches	X 25.4	=	millimeters (mm)	X	0.03937	=	inches
feet	X 0.3048	=	meters (m)	X	3.281	=	feet
yards	X 0.9144	=	meters (m)	X	1.0936	=	yards
miles	X 1.6093	=	kilometers (km)	X	0.6214	=	miles
inches	X 2.54	=	centimeters (cm)	X	0.3937	=	inches
microinches	X 0.0254	=	micrometers (Mm)	X	39.37	=	microinches

AREA
inches^2	X 645.16	=	millimeters^2 (mm^2)	X 0.00155		=	inches^2
inches^2	X 6.452	=	centimeters^2 (cm^2)	X 0.155		=	inches^2
feet^2	X 0.0929	=	meters^2 (m^2)	X 10.764		=	feet^2
yards^2	X 0.8361	=	meters^2 (m^2)	X 1.196		=	yards^2
acres	X 0.4047	=	hectacres (10^{4}m^2) (ha)	X 2.471		=	acres
miles^2	X 2.590	=	kilometers^2 (km^2)	X 0.3861		=	miles^2

VOLUME
inches^3	X 16387	=	millimeters^3 (mm^3)	X	0.000061	=	inches^3
inches^3	X 16.387	=	centimeters^3 (cm^3)	X	0.06102	=	inches^3
inches^3	X 0.01639	=	liters (l)	X	61.024	=	inches^3
quarts	X 0.94635	=	liters (l)	X	1.0567	=	quarts
gallons	X 3.7854	=	liters (l)	X	0.2642	=	gallons
feet^3	X 28.317	=	liters (l)	X	0.03531	=	feet^3
feet^3	X 0.02832	=	meters^3 (m^3)	X	35,315	=	feet^3
fluid oz	X 29.57	=	milliliters (ml)	X	0.03381	=	fluid oz
yards^3	X 0.7646	=	meters^3 (m^3)	X	1.3080	=	yards^3
teaspoons	X 4.929	=	milliliters (ml)	X	0.2029	=	teaspoons
cups	X 0.2366	=	liters (l)	X	4.227	=	cups

MASS
ounces (av)	X 28.35	=	grams (g)	X	0.03527	=	ounces (av)
pounds (av)	X 0.4536	=	kilograms (kg)	X	2.2046	=	pounds (av)
tons (2000 lb)	X 907.18	=	kilograms (kg)	X	0.001102	=	tons (2000 lb)
tons (2000 lb)	X 0.90718	=	metric tons (t)	X	1.1023	=	tons (2000 lb)

FORCE
ounces–f(av)	X 0.278	=	newtons (N)	X	3.597	=	ounces–f(av)
pounds–f(av)	X 4.448	=	newtons (N)	X	0.2248	=	pounds–f(av)
kilograms–f	X 9.807	=	newtons (N)	X	0.10197	=	kilograms–f

TEMPERATURE

°F -40 0 32 40 80 98.6 120 160 200 212 240 280 320 °F
°C -40 -20 0 20 40 60 80 100 120 140 160 °C

Degrees Celsius (C) = 0.556 (F - 32)   Degree Fahrenheit (F) = 1.8C + 32

Multiply:	by:		to get:	Multiply:	by:		to get:
ACCELERATION							
feet/sec^2	X 0.3048	=	meters/sec^2 (m/s^2)	X 3.281		=	feet/sec^2
inches/sec	X 0.0254	=	meters/sec^2 (m/s^2)	X 39.37		=	inches/sec^2
ENERGY OR WORK (Watt-second = joule = newton-meter)							
foot-pounds	X 1.3558	=	joules (J)	X 0.7376		=	foot-pounds
calories	X 4.187	=	joules (J)	X 0.2388		=	calories
Btu	X 1055	=	joules (J)	X 0.000948		=	Btu
waft-hours	X 3600	=	joules (J)	X 0.0002778		=	watt-hours
kilowatt-hrs	X 3.600	=	megajoules (MJ)	X 0.2778		=	kilowatt-hrs
FUEL ECONOMY & FUEL CONSUMPTION							
miles/gal	X 0.42514	=	kilometers/liter (km/l)	X 2.3522		=	miles/gal

Note:
235.2/(mi/gal) = liters/100km
235.2/(liters/100km) = mi/gal

PRESSURE OR STRESS							
inches Hg (60F)	X 3.377	=	kilopascals (kPa)	X 0.2961		=	inches Hg
pounds/sq in.	X 6.895	=	kilopascals (kPa)	X 0.145		=	pounds/sq in
inches H$_2$0 (60F)	X 0.2488	=	kilopascals (kPa)	X 4.0193		=	inches H$_2$O
bars	X 100	=	kilopascals (kPa)	X 0.01		=	bars
pounds/sq ft	X 47.88	=	pascals (Pa)	X 0.02088		=	pounds/sq ft
POWER							
horsepower	X 0.746	=	kilowatts (kW)	X 1.34		=	horsepower
ft-lbf/min	X 0.0226	=	watts (W)	X 44.25		=	ft-lbf/min
TORQUE							
pound-inches	X 0.11298	=	newton-meters (N-m)	X 8.851		=	pound-inches
pound-feet	X 1.3558	=	newton-meters (N-m)	X 0.7376		=	pound-feet
pound-inches	X 0.0115	=	kilogram-meters (Kg-M)	X 87		=	pound-feet
pound-feet	X 0.138	=	kilogram-meters (Kg-M)	X 7.25		=	pound-feet
VELOCITY							
miles/hour	X 1.6093	=	kilometers/hour (km/h)	X 0.6214		=	miles/hour
feet/sec	X 0.3048	=	meters/sec (m/s)	X 3.281		=	feet/sec
kilometers/hr	X 0.27778	=	meters/sec (m/s)	X 3.600		=	kilometers/hr
miles/hour	X 0.4470	=	meters/sec (m/s)	X 2.237		=	miles/hour

COMMON METRIC PREFIXES

mega	(M)	= 1,000,000	or	10^6	centi	(c)	= 0.01	or	10^{-2}
kilo	(k)	= 1,000	or	10^3	milli	(m)	= 0.001	or	10^{-3}
hecto	(h)	= 100	or	10^2	micro	(u)	= 0.000,001	or	10^{-6}

Index

A
ARE, 99
Acetylene torch, 56
Alternator, 105
  mounting, 105
Anti-seize compound, 94

B
B.C. Gerolamy, 47, 48, 50, 59
BRE, 4, 5, 6
Balzer, Norm, 103,114
Bearing crush, 38
Bearing oil clearance, 38
  check, 75
Berry, Dick, 122
Billet crankshaft, 20
Blackburn, Logan, 6, 71
Block-deck surface, O-ring, 16
Bob Sharp Racing, 5, 6, 7, 8, 25, 29, 41, 105
Bottom-end lubrication, 15
  four-cylinder engines, 15
  six-cylinder engines, 15
Bowman, Keith, 109
Burette, 52

C
Caldwell, John, 7
Camshaft, asymmetrical, 64, 65
  street performance, 62
  symmetrical, 64
Camshaft drive, install, 81, 91
Camshaft-drive chain tensioner, 81
  install, 81, 92
  limiter, 81
  slotted, 81
Camshaft duration, 65
Camshaft lobe, 61
  centers, 65
  closing-ramp profile, 64, 65
  lift, 65, 82
  opening-ramp profile, 64, 65
  rubbing-pad squareness, check, 79
  shift, 69
  specification card, 81, 82
  wipe pattern, 69-70, 80, 92
Camshaft oiling internal, 15
  spray bar, 15, 100
Camshaft sprocket, 81, 83
  adjustable 92
  dowel, 80
  install, 91-92, 93
  offset bushing, 83
Camshaft tower alignment, 54, 88
  cast iron, 59
  install, 55
  shim, 55, 79
Camshaft thrust plate, install, 91
Camshaft timing, 80, 82
  check, 81
Carburetor jet sizing, 111
Carburetor size select 107
  throttle bore, 109
  venturi, 109
Carburetor-sizing formula, 111
Carburetor, Dellorto, 106,108
Carburetor, Mikuni, 106, 108, 109, 111-117
  accelerator pump, 112, 113-114
  adjust, 115
  choke valve, 114
  float, 111
  fuel level, 112-113
  idle circuit, 115
  idle adjustment, 115
  jet selection, 115
  jets, 115
  low-speed circuit, 115
  main system, 112
  homogeneous-type, 115
  independent type, 115
  pilot, 112
  starter, 112, 114
  synchronize, 115
  tuning, 117
Carburetor, Mikuni/Solex, see Carburetor, Mikuni
Carburetor, SU, 121-125
  air/fuel mixture, adjust, 123
  basic function, 122
  emulsion-tubes/jet blocks, 106
  float adjustment, 124
  inspection, 124
  operation, 125
  float-level inspection, 124, 125
  idle-air/fuel mixture, 123
  idle-airflow synchronization, 123
  idle jets, 106
  idle speed, 123
  jet-needle inspection, 124
  needle-and-seat operation, 125
  nozzle inspection, 124
  suction chamber, 122
  suction piston, 122,123
  suction-piston/chamber assembly, service, 123
  synchronize, 123
Carburetor, Weber, 106, 107, 108, 117-121
  accelerator system, 119-120
  accelerator pump, 119
  air-correction jet, 120
  choke cable, 121
  emulsion tube, 120
  float system, 117
  float drop, 118
  float level, 118
  fuel pressure, 118
  fuel volume, 118
  idle system, 118
  installation angle, 118
  low-speed system, 118
  main jet, 120
  main system, 120-121
  outer-venturi size, 120
  starter system, 121
Carkhuff, Dave, 3, 7
Carney, Frank, 101
Chain guide, 14
Charging system, 104
Charts
  Basic Engine Specifications, 4
  Bore/Stroke Combinations, 4
  Crankshaft Specifications, 21
  Cylinder-Block Specifications, 11
  Cylinder-head port flow 48, 49
  Engine Dynamometer Test Log, 108
  Engine-Assembly Torques, 86
  FIA Group-2 Cylinder Heads, 58
  Metric Customary-Unit Equivalents, 141-142
  Mikuni/Solex 40PHH 44PHH, 110
  Nissan Motorsports Connecting Rods, 39
  Pressure-Plate Comparison, 140
  Road-Racing Carburetor Applications, 109
  Sealing Rings, 17
  Spark plug cross-reference, 102
  Standard Torque--NMC Bolts, 86
  Standard Valves, 42
  Stock Nissan Cast Pistons, 27
  Stock Nissan Connecting Rods, 34
  90-Series SU Carburetor Needles, 124
Circlip, 31, 74
Clement, Morris, 9, 10
Clutch assembly, install, 95
Combustion chamber. 52
  shape,52
  volume, 41, 42
  adjust, 52
  measure, 52
  weld, 46, 55
Compression ratio, 42, 43
Connecting rod alignment, 34
  assemble, 86
  balance, 35-37
  Carrillo, 40
  Crower, 40
  factory optional, 39
  investment cast, 40
  lost-wax cast, 40
  Magnaflux, 34
  Mechart, 40
  mild prep 33
  polish, 35
  race prep, 35
  shot-peen, 3s, 37
  size, 36
Connecting-rod bearing check, 74
  bearing select, 75
  install, 86
Connecting-rod bolt, 37-38
  big-block Chevy, 37
  torque, 87
Connecting-rod bearing journal, 75
  center-to-center length, 34, 38, 77
  pin oiling, 35, 36
  selection, 33
  sizing, 34, 35
  small end, 38
Conner, Jim, 126
Coolant-flow restrictor, 95
Coolcase, 23
Core plug, 11, 12
  install, 84
Coykendall, Bill, 7
Crankcase-vent screen, install, 85
Crank-fire ignition XXX
  crank-fire-ignition paddle, 24
Crankshaft, 19-25
  counterweight tapered 20
  damper, 24, 25
  install, 93
  end play, 76
  flywheel flange, 20
  heat treat, 22, 23
  key, 93
  oil hole preparation, 20-25
  register, see Crankshaft flywheel flange rotation, check, 86
  runout, 20, 21, 24
  sprocket, 77
  straightness, 20
Crankshaft-bearing oil holes, enlarge, 76
Crankshaft-journal oil grooves, 22
Crowe, Gene, 7,102
Cylinder Head Abrasives, 47
Cylinder block, 71
  deck, 11, 13
  clean 12, 13
  interior paint, 16
  race prep, 12-18
  rebore, 12
Cylinder-lock interior, smooth, 13
Cylinder-block oil-passage plugs
  fit 71
  install, 85
Cylinder-block preparation, dry-sump applications, 18
Cylinder bore, 18
  finish, 18
  out-of-round, 72
  taper, 72
  trueness, 18
  valve relief, 15
Cylinder head, 41-60
  assemble, 78, 88-89
  bend, 54, 56
  clean, 54
  four valve, 59, 60
  install, 80, 89-91
  mill, 79
  port matching, 44
  race modifications, 45- 53
  repair, 53
  straighten, 56
Cylinder-head port bowl, 50
  finish, 50
  modify, 47-50
  mouth, 50
  short-side radius. 47
Cylinders, Siamesed. 10

D
Datsun Alley, 122
Deck height, 78
Degree wheel, 77
Devendorf, Don, 5, 6, 6, 9, 26, 28, 29, 38 95,129
Distributor, 101-104
  camshaft driven, 102
  dual-point, 101, 103
  electronic type, 103
  European, 103 Mallory 103
  race-engine applications, 103
  single-point, 101
  street application, 101 -103
Distributor drive, index, 93
Distributor point bounce, 102
Distributor reluctor, 102
Dry-sump oil pan, 99-100
Dry-sump pump, 100
  Aviad, 100
  Peterson, 100
  Weaver, 100
Dyson, Rob, 7

E
Electramotive, 5, 8,10, 22, 24, 28, 29, 36, 50, 65, 78,100,105,130
Electronic ignition, 102
Engine displacement, 46
Exhaust manifold, 126-130
  cast iron, 126
  Clifford Research, 126
  Cyclone, 127
  original-equipment, 126
  primary-pipe diameter, 126-128
  primary-pipe length, 127-128

Exhaust port
  match, 44
  modify, 47-50
  round, 47
  rectangular, 47
Exhaust-port liner, 46
Exhaust system, 126-130
  off-road racing, 126
  mild street, 126
  racing, 127
  road racing, 129
Exhaust-valve events, 82

F
FAR Performance, 5
FIA, 41
Fitzgerald, Jim, 6, 33, 47, 96
Flywheel install, 95
  low-inertia 24
Flywheel bolts, 20, 23
Flywheel dowels, 23, 24
  install, 24, 25
Freeze plug see Core plug Frellsen, Dave, 5
Fuel pressure. 112
Fuel-pressure regulator, 119

G
Group-2. 41

H
H-M Mold Repair, 53
Head bolts, torque 91
  L28 Turbo 91
Head gasket
  install, 91
  L-28 turbo, 12
  sealing-ring type, 91
Head-gasket O-rings, 87
  install, 89
Head-gasket fire ring, 42
Head-gasket thickness, 78
Hi-Tack, 90 91
How to Rebuild Your Nissan/Datsun OHC Engine, 8

I
Ignition system, crankfire 104
Inner spring-seat, machine 67
Intake port, D-shaped, 47
Intake-valve events, 82
Intake-valve seats, brass, 57
Intake/exhaust-manifold gasket
  fit, 83
  trim, 83
Iskenderian Racing Cams, 65, 81

K
Keeper-to-valve seal interference, check, 78
Knepp, John, 7
Koobation, John, 103

L
L.A. Billet Crankshaft Co., 22, 25
Lapping plate, 57
Lash pad, 61, 62
  install, 92
  select, 79
  thickness, 63, 70
Leary, Frank, 5
Leitzinger Racing, 6
Leitzinger Bob, 33,100
Link, Floyd, 7, 22
Long, Bill, 6
Low Spencer, 7, 61,104
Lubriplate, 85

M
M Main bearings align, 86
  check, 75
  install, 85
Main oil galley, 71
Main oil-galley plug
  remove, 13
  install 71, 84
Main-bearing bores, 13
Main-bearing caps, install, 85
Main-bearing-bore roundness, 13
Main-bearing-cap bolts, torque, 86
Malvern Racing, 109
Mason, Gerry, Jr., 6
Mason, Gerry,, Sr., 6 Meniscus, 52, 53
Moldex, 25
Morton, John, 5
Muffler, performance 126

N
Newman, Paul, 5, 7, 33, 41
Nitriding, 23

O
O-ring 91
  butting, 17
  install, 89
O.S. Giken, 59,60
Oil filter, 13
Oil-filter adapter, 97
Oil-gallery plug, 13, 77
  install, 84
Oil jet, 14,15
Oil pan external-pickup 97
  install, 94
Oil pressure, 97
Oil-pressure-sender hole, 13
Oil-pump block-off plate, install, 93
Oil pump, 96
  install, 93- 94
  modify, 97
Oil-pump pickup
Oil-pump pressure, adjust, 96
Oiling system four-cylinder, 14
  six-cylinder, 14
Omega Engineering, 56

P
Palmer, Norwin, 47
Parkinson, Dan, 5
Parts List 133-140
Pilot bushing, 21, 25
  install 85
Piston assembly, install, 87
Piston bore size, 29
Piston dome, 28, 29
Piston pin 29-30, 73
  clearance, 73
  diameter, 29
  floating, 30, 38, 74
  height, 29
  oiling, 29
  pressed-in, 30, 31
  racing, 30
  retainers 3t, 86
  retention 30, 31
Piston-pin buttons, 31
  Teflon, 30 74
Piston ring, 26, 28
  backside clearance, 72
  design, 31
  end gap, 32, 73
  position, 86
  install, 86
  material, 30
  side clearance, 72
  spreader, 87
  stepped scraper, 31
  types, 29

Piston-skirt oil hole, 36
Piston slap, 30
Piston-to-bore clearance, 18, 31, 72
Piston-to-cylinder-head clearance, 32, 77
Piston-to-valve clearance, 29, 32
  check, 83
Piston valve pocket 29
Piston wobble, 30
Pistons, 26-30
  Arias, 28, 29
  cast 27
  Cosworth 28
  domed racing 28
  forged racing 28
  Ross, 28, 29
  stock, 27, 28
  tapered, 31
  Venolia, 28, 29
Port-flow direction, 52
Ports round, 58
Pressure-relief valve, 97
Preston, Don, 6

Q
Quarter Master Industries. 95

R
Racing history, 4
Racing regulations, 45
Ray, John, 104
Relief valve, 96
Rickman, Mike, 7,19
Roberts, Dick, 5
Rocker arm, 61, 65, 69
  balance, 79
  lighten, 65
Rocker-arm contact-pad wipe pattern, check, 92
Rocker-arm-pivot, install, 88
Rocker-arm ratio, 65, 70, 82
  check, 65, 79
Rocker-arm weight, 65
Rosebud tip, 56

S
Sealing-ring head gasket, 16-18
Sharp, Bob, 5,128,129
Short-block mock-up, 77
Slack-side chain guide, install, 91
Slover s Porting Service, 48
Spark accuracy, 103
Spark plugs index, 94
  install, 94
Spirolox, 31, 74
Spray bar, 62
  install, 92
  kit, 63
Spring load, 66
Starter 104
  direct drive, 104
  gear reduction, 104
Street engine, supercharged 44
Street engine, turbocharged, 44,126
Supplier List, 131-132

T
TC (Top Center), find, 77
Temperature indicator, 56
Tempilstik, 56
Tension-side chain guide, install, 91
Thermostat-housing, install, 94
Threaded holes, chamfer, 14, 77
Tilton Engineering, 24
Timing chain, 81
  bright link, 81, 91
  guides 81
  install, 91
  slack-side-guide alignment, 81
  tensioner, 81

Tuftride, 23
Turbo Toms, 127
Turbo block, 10
Turbo muffler, 126
Turbocharging, 129
Twin-choke sidedraft carburetors, 106

U
Uni-Syn, 115,123

V
Valve, unshroud, 15, 52
Valve back-cutting, 51
Valve depth, 51
Valve guide
  bronze, 57
  cast iron, 57
  replace, 57
Valve-head margin, 51
Valve-head shape, 51
Valve job, 50-51
  three-angle, 50
Valve keeper, 67
Valve lash, 82
Valve-lash pad, 67-69, 79
  thickness, 80
Valve lift, 82
Valve-lift interference, check, 68, 69
Valve relief, 15
Valve seat, 50- 51
  bottom cut, 50
  radius, 50
  replace, 57
  top cut, 50
Valve-seat shaping, 51
Valve spring, 65- 67, 78
  coil bind, 62, 76
  compressor, 88
  height, 83
  installed height, 78, 79
  measure, 79
  load, 62 66, 78
Valve-spring pressure, 62
Valve-spring rate, 78
Valve-spring retainer, 62, 67
  aluminum, 64, 67
  small-block Ford, 67
  steel, 64, 67
  titanium, 64
Valve-spring seat, 79
Valve-spring shim, 68, 78, 79
  install, 88
Valve-spring step, 68
Valve-spring-seat cutter, 68
Valve-stem seal, 68
Valve-stem-tip height, 79
Valve timing, check, 82
Valve-tip wear cap, 66
Valve train, direct oiling, 62
Valve-train instability, 64, 66, 67
Valve-train lubrication, 62
Valves steel, 66
  titanium, 66
  unshroud, 45
Velocity stack, 109

W
Welding, repair, 54
Wet-sump lubrication, 97
Wet-sump oil pan, 98, 99
Wyatt. Tom. 111, 127

Z
ZX-Turbo high-volume oil pump. 96